Towards Greenhouse Gas Mitigation: Novelty in Heterogeneous Catalysis

Towards Greenhouse Gas Mitigation: Novelty in Heterogeneous Catalysis

Editor

Wasim Ullah Khan

MDPI • Basel • Beijing • Wuhan • Barcelona • Belgrade • Manchester • Tokyo • Cluj • Tianjin

Editor
Wasim Ullah Khan
Interdisciplinary Research
Centre for Refining &
Advanced Chemicals,
Research Institute, King Fahd
University of Petroleum &
Minerals
Dhahran 31261, Saudi Arabia

Editorial Office
MDPI
St. Alban-Anlage 66
4052 Basel, Switzerland

This is a reprint of articles from the Special Issue published online in the open access journal *Energies* (ISSN 1996-1073) (available at: https://www.mdpi.com/journal/energies/special_issues/Towards_Greenhouse_Gas_Mitigation_Novelty_in_Heterogeneous_Catalysis).

For citation purposes, cite each article independently as indicated on the article page online and as indicated below:

LastName, A.A.; LastName, B.B.; LastName, C.C. Article Title. *Journal Name* **Year**, *Volume Number*, Page Range.

ISBN 978-3-0365-4445-8 (Hbk)
ISBN 978-3-0365-4446-5 (PDF)

© 2022 by the authors. Articles in this book are Open Access and distributed under the Creative Commons Attribution (CC BY) license, which allows users to download, copy and build upon published articles, as long as the author and publisher are properly credited, which ensures maximum dissemination and a wider impact of our publications.

The book as a whole is distributed by MDPI under the terms and conditions of the Creative Commons license CC BY-NC-ND.

Contents

About the Editor vii

Preface to "Towards Greenhouse Gas Mitigation: Novelty in Heterogeneous Catalysis" ix

Wasim Ullah Khan
Energy and Environment—Towards Greenhouse Gas Mitigation: Novelty in Heterogeneous Catalysis
Reprinted from: *Energies* **2022**, *15*, 3795, doi:10.3390/en15103795 1

Ahmed Abasaeed, Samsudeen Kasim, Wasim Khan, Mahmud Sofiu, Ahmed Ibrahim, Anis Fakeeha and Ahmed Al-Fatesh
Hydrogen Yield from CO_2 Reforming of Methane: Impact of La_2O_3 Doping on Supported Ni Catalysts
Reprinted from: *Energies* **2021**, *14*, 2412, doi:10.3390/en14092412 5

Naushad Ahmad, Fahad Alharthi, Manawwer Alam, Rizwan Wahab, Salim Manoharadas and Basel Alrayes
Syngas Production via CO_2 Reforming of Methane over $SrNiO_3$ and $CeNiO_3$ Perovskites
Reprinted from: *Energies* **2021**, *14*, 2928, doi:10.3390/en14102928 19

V. P. Singh, Mirgender Kumar, Moolchand Sharma, Deepika Mishra, Kwang-Su Seong, Si-Hyun Park and Rahul Vaish
Synthesis of BiF_3 and BiF_3-Added Plaster of Paris Composites for Photocatalytic Applications
Reprinted from: *Energies* **2021**, *14*, 5159, doi:10.3390/en14165159 31

Wasim Ullah Khan, Mohammad Rizwan Khan, Rosa Busquets and Naushad Ahmad
Contribution of Oxide Supports in Nickel-Based Catalytic Elimination of Greenhouse Gases and Generation of Syngas
Reprinted from: *Energies* **2021**, *14*, 7324, doi:10.3390/en14217324 45

Kenji Taira and Reiko Murao
High Dispersion of CeO_2 on CeO_2/MgO Prepared under Dry Conditions and Its Improved Redox Properties
Reprinted from: *Energies* **2021**, *14*, 7922, doi:10.3390/en14237922 57

Daniel Lach, Jaroslaw Polanski and Maciej Kapkowski
CO_2—A Crisis or Novel Functionalization Opportunity?
Reprinted from: *Energies* **2022**, *15*, 1617, doi:10.3390/en15051617 75

Maciej Kapkowski, Tomasz Siudyga, Piotr Bartczak, Maciej Zubko, Rafal Sitko, Jacek Szade, Katarzyna Balin, Bartłomiej S. Witkowski, Monika Ożga, Rafał Pietruszka, Marek Godlewski and Jaroslaw Polanski
Catalytic Removal of NOx on Ceramic Foam-Supported ZnO and TiO_2 Nanorods Ornamented with W and V Oxides
Reprinted from: *Energies* **2022**, *15*, 1798, doi:10.3390/en15051798 95

About the Editor

Wasim Ullah Khan

Assistant Professor (Research) Dr. Wasim Ullah Khan received his BE in Chemical Engineering from the University of Engineering and Technology Lahore in 2008. He obtained his MS degree in Chemical Engineering from King Saud University Riyadh, Saudi Arabia, in 2015. He then moved to New Zealand to continue his doctoral degree under the supervision of Dr. Alex Yip in the Energy and Environmental Catalysis Group at the University of Canterbury, New Zealand. He is currently serving as an Assistant Professor (Research) at the Interdisciplinary Research Center for Refining & Advanced Chemicals, Research Institute, King Fahd University of Petroleum & Minerals, Dhahran, Saudi Arabia where his research focuses on the synthesis, characterization, and catalytic properties of bimetallic catalysts for hydrogen production from ammonia decomposition and design and synthesis of transition metal-based catalyst for in situ copolymerization of ethylene with polar monomer.

Preface to "Towards Greenhouse Gas Mitigation: Novelty in Heterogeneous Catalysis"

Climate change caused by greenhouse gas emissions has encouraged the scientific community to find ways to mitigate these environmentally detrimental gases. The conversion of major greenhouse gases, including methane and carbon dioxide, with the utilization of catalysts in general and heterogeneous catalysts in particular has made significant progress in recent decades. The wider scientific community of MDPI can benefit from this Special Issue in terms of understanding the basic research involved in reaction engineering and catalysis. Finally, I, as the guest editor of this Special Issue, would like to extend my appreciation to MDPI and the Energies team for providing this exciting opportunity of learning and growth as well as the editorial staff, especially Ms. Marija Belic, for their continuous support and consideration. I must acknowledge the fact that such interactions are an excellent platform for young researchers for their scientific growth and I hope the readers enjoy this piece of research.

Wasim Ullah Khan
Editor

Editorial

Energy and Environment—Towards Greenhouse Gas Mitigation: Novelty in Heterogeneous Catalysis

Wasim Ullah Khan [1,2]

[1] Department of Chemical and Process Engineering, University of Canterbury, 20 Kirkwood Avenue, Upper Riccarton, Christchurch 8041, New Zealand; wasimkhan49@gmail.com or wasim.khan@kfupm.edu.sa

[2] Interdisciplinary Research Centre for Refining & Advanced Chemicals, Research Institute, King Fahd University of Petroleum & Minerals, Dhahran 31261, Saudi Arabia

Climate change, a consequence of global warming, is a global issue resulting due to greenhouse gas (GHG) emissions. The main GHGs of concern are carbon dioxide (CO_2), methane, and nitrogen oxide. Scientists have utilized technologies in recent decades to mitigate GHGs. Among the proposed technological solutions, catalysis—and in particular heterogeneous catalysis—has played a vital role in the abatement of GHGs.

One of the strategies to combat methane emissions is combustion, and catalytic combustion offers economic benefits due to the fact that the lower methane concentration in combustion emissions is less harmful than by-products such as formaldehyde. Nanotechnological advancement in heterogeneous catalysis for combustion has significantly reformed the process. The catalytic conversion of carbon dioxide is carried out via different routes, hence suggesting the economic development of energy-efficient catalytic CO_2 conversion to useful products. Some of the well-known catalytic routes to convert methane and CO_2 include steam methane reforming, dry reforming of methane, partial oxidation of methane, methane decomposition, reverse water gas shift, CO_2 hydrogenation to methanol, and CO_2 hydrogenation to higher alcohols.

The investigation of metal/metal oxide nanoparticles anchored onto an oxide support in a heterogeneous catalysis is significantly important for understanding the nature and extent of the metal–support interaction which affects the catalytic activity and product selectivity. The preparation of a heterogeneous catalyst involves different elevated temperature steps, including oxidation and reduction, which influence the morphology of the catalyst. The metal–support interaction also causes morphological changes such as alloy formation, sintering of metal particles, inter-diffusion, and encapsulation. Therefore, the preparation steps need to be optimized to obtain well-dispersed metal nanoparticles anchored onto the oxide support. The recent developments in spectroscopic and microscopic characterization techniques, as well as density functional theory, have facilitated scientists in predicting the performance of the catalysts and proposing hypotheses before the reactions. Later, the suggested hypotheses are validated by characterizing the catalysts after the reactions.

Considering the above-mentioned developments in heterogeneous catalysis for GHG abatement, this Special Issue is mainly focused on the novel advancements in heterogeneous catalysis for the mitigation of GHGs. Potential topics include, but were not limited to the following:

- Heterogeneous catalysts for steam methane reforming;
- Dry methane reforming, partial oxidation of methane;
- Methane decomposition;
- Reverse water–gas shift;
- CO_2 hydrogenation to methanol;
- CO_2 hydrogenation to higher alcohols.

Considering the COVID-19 crisis, the response to our call was excellent, with the following statistics:

Citation: Khan, W.U. Energy and Environment—Towards Greenhouse Gas Mitigation: Novelty in Heterogeneous Catalysis. *Energies* **2022**, *15*, 3795. https://doi.org/10.3390/en15103795

Received: 19 May 2022
Accepted: 20 May 2022
Published: 21 May 2022

Publisher's Note: MDPI stays neutral with regard to jurisdictional claims in published maps and institutional affiliations.

Copyright: © 2022 by the author. Licensee MDPI, Basel, Switzerland. This article is an open access article distributed under the terms and conditions of the Creative Commons Attribution (CC BY) license (https://creativecommons.org/licenses/by/4.0/).

- Submissions: (8);
- Publications: (7);
- Rejections/withdrawals: (1);
- Article types: review article (1); research articles (6).

A brief overview of the contributions in this Special Issue:

Abasaeed et al. [1] submitted the very first article to the SI and demonstrated the modification of support materials such as alumina (Al_2O_3) and zirconia (ZrO_2) by doping lanthanum oxide (La_2O_3) in the range from 0 to 20 wt%. These modified support-based nickel catalysts were further investigated for a dry methane reforming reaction. The catalytic performance tests revealed that lanthanum oxide modification positively influenced the conversions and, depending upon the base support, modified catalysts outperformed the unmodified catalysts. The improvement in surface basic properties, specific surface area, metal dispersion, and lower reduction temperatures of modified catalysts were the major factors behind their superior performance during the dry reforming reaction.

The role of strontium (Sr) and cerium (Ce) in nickel-based perovskite catalysts during methane reforming using carbon dioxide was studied by Ahmed et al. [2]. The catalyst characterization results showed that a $CeNiO_3$ perovskite catalyst exhibited a higher number of reducible species, BET surface area, pore volume, and nickel dispersion as compared to a $SrNiO_3$ catalyst. These factors had a significant impact on catalytic activity results and the $CeNiO_3$ catalyst obviously outperformed the $SrNiO_3$ catalyst. The catalyst stability tests also followed the same trend and $SrNiO_3$ deactivated more than $CeNiO_3$, which was assigned to carbon deposition.

Singh et al. [3] investigated the photocatalytic performance of BiF_3 and Bi_2O_3 and found that BiF_3 presented superior activity. Considering the relatively better photocatalytic performance of BiF_3, they incorporated it in plaster of Paris (POP) with varying amounts of BiF_3 between 0 and 10 wt%. The photocatalytic activity of BiF_3-modified POP evaluated using Resazurin (Rz) ink under ultraviolet (UV) light irradiation demonstrated that an increase in the amount of BiF_3 improved ink removal and, hence, enhanced photocatalytic activity. This was further substantiated with UV visible spectroscopy which quantified the rate of de-coloration of modified POP samples.

Khan et al. [4] evaluated the performance of various supports for nickel-based catalysts during the dry reforming of methane. The exploration of supports, such as H-ZSM-5 zeolite, Y-zeolite, and alumina (Al_2O_3), under dry reforming conditions indicated that an alumina-supported nickel catalyst exhibited higher catalytic activity than that of zeolite-supported catalysts. On the contrary, the alumina-supported nickel catalyst deactivated more than the zeolite-supported catalysts and, hence, zeolite-supported catalysts showed more stability, especially with just 2% deactivation, and the H-ZSM-5-supported nickel catalyst remained the most stable catalyst.

Taira and Murao [5] studied the role of ceria (CeO_2) dispersion in CeO_2/MgO synthesized under a dry condition and its improved redox properties during a methane dry reforming reaction. The impregnation of ceria over MgO under a dry condition maintained the ceria particle diameter at approximately 3 nm, even at 800 °C, and a slight agglomeration of ceria particles (~5 nm) was noted when impregnated under ambient conditions. The catalyst characterization results identified that dry condition impregnation imparted a higher number of mobile oxygen species than the ambient condition counterpart, which played its positive role during the dry reforming reaction.

This SI also witnessed a review article by Lach and his teammates [6] focused on comparing the perspectives regarding carbon dioxide as a crisis or an opportunity and working towards useful utilization. The review highlighted the carbon dioxide emissions contributing to global warming and a possible solution for its abatement by using carbon dioxide as a raw material, such as in carbon dioxide methanation, which is one of the starting points for successive projects such as synthesis, polymers, and/or fuels, and power-to-gas applications. Hence, the review article discussed the carbon dioxide methanation reaction and development and progress in catalyst design for this reaction.

Finally, an article was published by Kapkowski et al. [7] to demonstrate the role of zinc oxide (ZnO) and titania (TiO$_2$) nanorods supported on ceramic foam (using alumina, silicon carbide, and zirconia substrates) during catalytic NOx removal. They also explored the ornaments of ceramic foam containing ZnO and TiO$_2$ nanofilaments with oxides of vanadium (V) and tungsten (W). The catalytic activity results revealed that among alumina-based form-supported TiO$_2$ nanorods ornamented with oxides of V and W, 1% W/TiO$_2$/Al$_2$O$_3$ outperformed the rest of the catalysts, whereas in the case of ZnO nanorods, 1% V,W(3:7)/ZnO/Al$_2$O$_3$ exhibited the highest activity. Using 1% loading of V and/or W as optimum, ZrO$_2$-supported ZnO nanorods (1% W,V/ZnO/ZrO$_2$) were found to have the highest conversion.

Funding: This research received no external funding.

Conflicts of Interest: The author declares no conflict of interest.

References

1. Abasaeed, A.; Kasim, S.; Khan, W.; Sofiu, M.; Ibrahim, A.; Fakeeha, A.; Al-Fatesh, A. Hydrogen Yield from CO$_2$ Reforming of Methane: Impact of La$_2$O$_3$ Doping on Supported Ni Catalysts. *Energies* **2021**, *14*, 2412. [CrossRef]
2. Ahmad, N.; Alharthi, F.; Alam, M.; Wahab, R.; Manoharadas, S.; Alrayes, B. Syngas Production via CO$_2$ Reforming of Methane over SrNiO$_3$ and CeNiO$_3$ Perovskites. *Energies* **2021**, *14*, 2928. [CrossRef]
3. Singh, V.P.; Kumar, M.; Sharma, M.; Mishra, D.; Seong, K.-S.; Park, S.-H.; Vaish, R. Synthesis of BiF$_3$ and BiF$_3$-Added Plaster of Paris Composites for Photocatalytic Applications. *Energies* **2021**, *14*, 5159. [CrossRef]
4. Khan, W.U.; Khan, M.R.; Busquets, R.; Ahmad, N. Contribution of Oxide Supports in Nickel-Based Catalytic Elimination of Greenhouse Gases and Generation of Syngas. *Energies* **2021**, *14*, 7324. [CrossRef]
5. Taira, K.; Murao, R. High Dispersion of CeO$_2$ on CeO$_2$/MgO Prepared under Dry Conditions and Its Improved Redox Properties. *Energies* **2021**, *14*, 7922. [CrossRef]
6. Lach, D.; Polanski, J.; Kapkowski, M. CO$_2$—A Crisis or Novel Functionalization Opportunity? *Energies* **2022**, *15*, 1617. [CrossRef]
7. Kapkowski, M.; Siudyga, T.; Bartczak, P.; Zubko, M.; Sitko, R.; Szade, J.; Balin, K.; Witkowski, B.S.; Ożga, M.; Pietruszka, R.; et al. Catalytic Removal of NOx on Ceramic Foam-Supported ZnO and TiO$_2$ Nanorods Ornamented with W and V Oxides. *Energies* **2022**, *15*, 1798. [CrossRef]

Article

Hydrogen Yield from CO$_2$ Reforming of Methane: Impact of La$_2$O$_3$ Doping on Supported Ni Catalysts

Ahmed Abasaeed [1], Samsudeen Kasim [1], Wasim Khan [1,2,*], Mahmud Sofiu [1], Ahmed Ibrahim [1], Anis Fakeeha [1,3] and Ahmed Al-Fatesh [1]

[1] Chemical Engineering Department, College of Engineering, King Saud University, P.O. Box 800, Riyadh 11421, Saudi Arabia; abasaeed@ksu.edu.sa (A.A.); sofkolajide2@gmail.com (S.K.); mahmudsofiu@gmail.com (M.S.); aididwthts2011@gmail.com (A.I.); anishf@ksu.edu.sa (A.F.); aalfatesh@ksu.edu.sa (A.A.-F.)
[2] Department of Chemical and Process Engineering, University of Canterbury, 20 Kirkwood Avenue, Upper Riccarton, Christchurch 8041, New Zealand
[3] King Abdullah City for Atomic & Renewable Energy, Energy Research & Innovation Center (ERIC) in Riyadh, Riyadh 11451, Saudi Arabia
* Correspondence: wasimkhan49@gmail.com

Abstract: Development of a transition metal based catalyst aiming at concomitant high activity and stability attributed to distinguished catalytic characteristics is considered as the bottleneck for dry reforming of methane (DRM). This work highlights the role of modifying zirconia (ZrO$_2$) and alumina (Al$_2$O$_3$) supported nickel based catalysts using lanthanum oxide (La$_2$O$_3$) varying from 0 to 20 wt% during dry reforming of methane. The mesoporous catalysts with improved BET surface areas, improved dispersion, relatively lower reduction temperatures and enhanced surface basicity are identified after La$_2$O$_3$ doping. These factors have influenced the catalytic activity and higher hydrogen yields are found for La$_2$O$_3$ modified catalysts as compared to base catalysts (5 wt% Ni-ZrO$_2$ and 5 wt% Ni-Al$_2$O$_3$). Post-reaction characterizations such as TGA have showed less coke formation over La$_2$O$_3$ modified samples. Raman spectra indicates decreased graphitization for La$_2$O$_3$ catalysts. The 5Ni-10La$_2$O$_3$-ZrO$_2$ catalyst produced 80% hydrogen yields, 25% more than that of 5Ni-ZrO$_2$. 5Ni-15La$_2$O$_3$-Al$_2$O$_3$ gave 84% hydrogen yields, 8% higher than that of 5Ni-Al$_2$O$_3$. Higher CO$_2$ activity improved the surface carbon oxidation rate. From the study, the extent of La$_2$O$_3$ loading is dependent on the type of oxide support.

Keywords: Al$_2$O$_3$; CO$_2$ reforming; La$_2$O$_3$; CH$_4$; ZrO$_2$

1. Introduction

The decrease of fossil fuel energy and the dilemma of environmental pollution urged a large number of researchers to maximize the conversions of methane and carbon dioxide into useful products such as hydrogen. Hydrogen is a benign source of energy. It is mainly obtained from biomass pyrolysis and thermal reforming. Methane, the main component of natural gas, can be obtained from various resources like shale gas and the fracking process, which has increased the availability of natural gas from infrequent deposits [1,2]. Moreover, the utilization of biogas is gaining momentum in recent years [3,4]. In the field of heterogeneous catalysis, particularly, in the latest decades, dry reforming of methane is regarded as one of the best prospective ways of conversion [5–7]. However, the dry reforming reaction as shown in Equation (1) is highly endothermic and thus requires high reaction temperatures. The process produces synthesis gas that has an appropriate ratio of H$_2$ to CO suitable for Fischer–Tropsch synthesis [8]. Steam reforming of methane remains the best available industrial process for generating synthesis gas [9,10]. The requirement and the utilization of synthesis gas production are continuously increasing [11]. During methane dry reforming (DRM), CO$_2$ is employed as an oxidant, which draws the interest and the likelihood of seizing and recycling CO_2 from the exhaust flue gases of industrial and power plants. DRM is presently not industrially applied because of the heavy coking and sintering of the catalysts in high-temperature reforming reactions [12]. Thus, it is essential to find an innovative catalyst that endures sintering and coking. The sintering of the metallic phase and carbon formation on the surface of the catalyst, which causes the deactivation, originate from operating conditions that facilitate the side reactions, which involve the cracking of the CH_4 Equation (2) and the reverse Boudouard reaction Equation (3) resulting from the combinations of CO [13].

$$CH_4 + CO_2 \rightarrow 2H_2 + 2CO \tag{1}$$

$$CH_4 \rightarrow 2H_2 + C \tag{2}$$

$$CO_2 + C \rightarrow CO + CO \tag{3}$$

H_2 yield through DRM is significantly affected by dissociation of CH_4 over Ni-supported surface and the gasification of carbon formed by CO_2. The reverse water gas shift reaction and the action of the H_2 spillover on the surface affect the H_2 yield substantially. The hydrogen spill over enhances hydroxyl formation and catalytic activity toward CO oxidation at the metal/oxide interface. The hydroxyl groups at the metal/support interface react with CO to produce CO_2. Similarly, the reverse water gas shift (CO_2 + $H_2O \rightarrow CO + H_2$) reduces the hydrogen; hence, the two phenomena involve the depletion of hydrogen and in turn influence the hydrogen yield. Noble metals like Pd, Ir, Pt, and Rh provide the extremely good performance of activity and stability but they are rare and expensive [14,15]. Transition metals like Ni are suitable alternatives for this reaction as they are stable and environmentally friendly [16,17]. Nonetheless, Ni-based catalysts are hampered by poor activity due to coke formations and sintering [18,19]. Hence, the major challenge is to come up with Ni catalysts capable of resisting carbon formation and sintering. Carbon generation may be opposed by controlling the reaction kinetics using suitable catalysts with proper constituents and supports. It has been established that metal–support interactions can alter both catalyst activity and activity maintenance. Bradford and Vannice had shown that support decoration of metal particle surfaces shattered big ensembles of metal atoms that served as active sites for carbon deposition and the sites in the metal–support interfacial region enhanced catalyst activity [20]. The choice of support type plays a vital role in DRM. Supports that possess O_2 species at the surface of the catalyst help the carbon oxidation on the metal [21,22]. The balance between the rate of methane decomposition and the rate of carbon gasification regulates the catalyst stability [23]. The preparation of mesoporous oxides, like Al_2O_3, possessing high surface area and precise pores have revealed motivating results to disperse the metal over the support structure [24–26]. Bian et al., in their investigation of Ni supported home-made mesoporous alumina in methane dry reforming, indicated that the formation of $NiAl_2O_4$ spinel is advantageous to activity and stability towards DRM reaction [27].

Newnham et al. synthesized nanostructured Ni-incorporated mesoporous alumina with various Ni loadings by hydrothermal method and tested them as catalysts for CO_2 reforming of methane [28]. The result displayed excellent stability when 10%Ni was used due to the fact that the Ni nanoparticles in these catalysts being highly stable towards migration/sintering under the reaction conditions. The presence of strong Ni–support interaction and/or active metal particles being confined to the mesoporous channels of the support. Al-Fatesh et al. studied dry reforming of methane using a series of nickel-based catalysts supported on γ-alumina promoted by B, Si, Ti, Zr, Mo, and W [5]. They concluded that the promoters enhanced the interaction between NiO and γ-alumina support and, hence, Ni dispersion and stability. On the other hand, ZrO_2 is prominent support with high thermal stability able to go through alteration in their acid–basic sites [29]. Numerous studies have displayed that the presence of ZrO_2 improves the thermal resistance, redox properties, oxygen storage capacity and gasification of deposited carbon [30,31]. Hu et al. examined the dry reforming of methane over Ni/ZrO_2 catalysts prepared via decomposition of nickel precursor under the influence of dielectric barrier discharge (DBD) plasma at ~150 °C [32]. It was found to improve activities due to the exposition of Ni (111) facets, smaller metal particles, and more tetragonal zirconia with increased oxygen vacancies. In the course of dry reforming of methane, oxygen species over the catalyst surface affected the catalytic performance and carbon deposition. Zhang et al. studied the effects of the surface adsorbed oxygen species tuned by doping with metals like La, Ce, Sm, and Y on the catalytic behavior [33]. Their results confirmed that the surface adsorbed oxygen species promoted both CO_2 activation and CH_4 dissociation. Doping La_2O_3 in supported Ni catalysts favor the CO_2 adsorption on the surface of the catalyst [34], alters the chemical and electronic state of Ni at the interface with the support and decreases the chemical interaction between Ni and the support causing the intensification of reducibility and higher dispersion of nickel [35]. Tran et al. studied the enhancement of La_2O_3 in the physicochemical features of cobalt supported over alumina for DRM using different temperatures and feed compositions [36]. Their results displayed that the La_2O_3 improved the H_2 activation; enriched oxygen vacancy and lowered the apparent activation energy of CH_4 consumption. The work of Lui et al. elaborated the promotional effects of La, Al, and Mn on Fe-modified clay supported by Ni catalysts used for dry reforming of methane [37]. The result of adding La, Al, and Mn altered the surface area, the basicity of the catalysts, and produced a smaller metallic Ni size. Moreover, Yabe et al. performed dry reforming of methane using several transition metals supported on ZrO_2 catalysts [38]. Their results exhibited high activity and low carbon deposition upon using 1 wt%Ni/10 mol%La-ZrO_2 catalysts.

In the present work, we assess the effect of the lanthanum oxide as a textual promoter of alumina and zirconia supports over Ni catalysts in the catalytic reforming of CH_4 with CO_2. The impact of different loadings of lanthanum oxide will be examined and their influence on the hydrogen yield. The output data will be further associated with the characterization results of BET, XRD, TPR, TEM, TPD, and TGA before and after the reaction. The difference in the basic supports originating from alumina and zirconia in terms of sintering and coking will be explored.

2. Materials and Methods

2.1. Materials

Mesoporous γ-Al_2O_3 (purity 99.99%, purchased from Norton Co (New York, NY, USA)), precursor of zirconium oxide ($ZrOCl_2 \cdot 8H_2O$, purity >99.0%, purchased from Fluka Chemika (Washington, DC, USA)), precursors of La_2O_3 and Ni, respectively, $(La(NO_3)_3 \cdot 6H_2O$ and $Ni(NO_3)_2 \cdot 6H_2O$ purchased from Sigma Aldrich (St. Louis, MO, USA)), double distilled water.

2.2. Catalyst Preparation

In the case of zirconium-based support (for 5Ni-10La-ZrO_2), the required amounts of zirconia (2.62 g) in the form of $ZrOCl_2 \cdot 8H_2O$ was ground completely and poured into the empty crucible. Then, the desired weights of $La(NO_3)_3 \cdot 6H_2O$ (0.265 g) and that of $Ni(NO_3)_2 \cdot 6H_2O$ (0.25 g) were added to the crucible containing the support to get a powder mixture. The mixture was ground well in the crucible to obtain a homogenous mixture. Purified water was poured slowly to the mixture to produce a paste while mixing. The paste was set to evaporate under room temperature condition until it dried. Thereafter, the dried sample was calcined at 700 °C for 3 h. The obtained catalysts were denoted as 5Ni-ZrO_2 for 5 wt% Ni supported over ZrO_2, and 5Ni-xLa-ZrO_2, where x = 10, 15, 20 wt%.

In the case of alumina-based support, the above-mentioned procedure was adopted by replacing zirconia with the required amounts of mesoporous γ-Al_2O_3. The obtained catalysts were denoted as 5Ni-Al_2O_3 for 5 wt% Ni supported over Al_2O_3 and 5Ni-xLa-Al_2O_3, where x = 10, 15, 20 wt%.

2.3. Catalyst Characterization and Activity

The catalyst activity test and characterization are described in detail in the supplementary information.

3. Results and Discussion

The surface texture of the catalysts was assessed via the nitrogen adsorption–desorption isotherms. Figure 1A shows the nitrogen adsorption isotherms of the fresh catalysts (5Ni-xLa$_2$O$_3$-ZrO$_2$, x = 0, 10, 15, and 20 wt%) and according to IUPAC labelling, catalysts are showing type IV isotherm with capillary condensation appearing at a relative pressure below the saturation pressure, and H1 hysteresis loop. These features are associated with mesoporous materials that have cylindrical pore geometry with narrow size distribution as well as relatively high uniformity [39]. Figure 1B exhibits the nitrogen adsorption isotherms of the fresh catalysts (5Ni-x%La-Al$_2$O$_3$). All samples indicate typical type IV adsorption/desorption isotherms with H1 hysteresis loop. Point of inflection at a relative pressure in the range of 0.6–0.75 corresponds to the capillary condensation which indicates the uniformity of the pores in mesoporous material [40,41]. The textural properties of the catalyst are given in Table S1 of the supplementary.

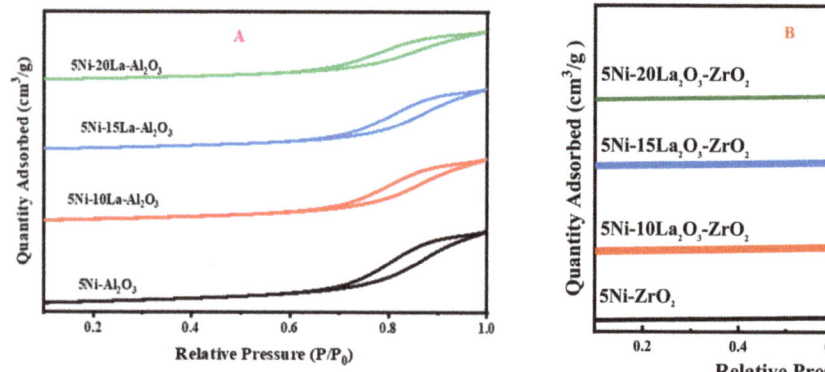

Figure 1. N$_2$ adsorption–desorption isotherms for fresh (**A**) 5Ni-xLa$_2$O$_3$+ZrO$_2$ and (**B**) 5Ni-x La$_2$O$_3$+Al$_2$O$_3$ catalyst calcined at 700 °C, (x = 0, 10, 15, and 20 wt%).

The reducibility of 5%Ni-x% La$_2$O$_3$-ZrO$_2$ and 5%Ni-x% La$_2$O$_3$-Al$_2$O$_3$ (x = 0, 10, 15, and 20) catalysts were examined by TPR and the patterns are shown in Figure 2. For the 5%Ni-x% La$_2$O$_3$-ZrO$_2$ catalysts (Figure 2A), two prominent reduction maxima are detected over the entire temperature range, which may be attributed to the reduction of different NiO species (NiO→Ni^0). The first reduction peak situated in region I at Tmax = 298 °C could be ascribed to the reduction of free NiO that is not attached to the support, and hence reduces easily at low temperature. Further, the peak in the region II at Tmax = 468 °C is allotted to the NiO reduction, attached to ZrO$_2$ by a moderately strong link and its reduction requires higher thermal energy. After support modification by means of La$_2$O$_3$, substantial variations in reduction kinetics were detected. Reduction maxima illustrating NiO reduction for the higher temperature peaks has shifted towards lower temperatures [42]. In the case of Figure 2B, the 5% Ni-x% La$_2$O$_3$-Al$_2$O$_3$ catalysts (x = 0, 10, 15, and 20) display a single peak in region III at higher temperatures. The un-promoted 5%Ni-Al$_2$O$_3$ catalyst shows a broad peak at 750 °C indicating that the reduced NiO is strongly attached to the support. When the support is modified with the addition of different loadings of La$_2$O$_3$, peaks of relatively lower areas appear at high temperatures. The La$_2$O$_3$ modified support catalyst is shifted to higher temperatures [43]. The highest La$_2$O$_3$ loading catalyst gives the highest peak shift. This means that the addition of La$_2$O$_3$ increases further the interaction between the NiO and the modified support. The quantitative analysis of H_2 consumption during H_2-TPR is displayed in Table S2 of the supplementary.

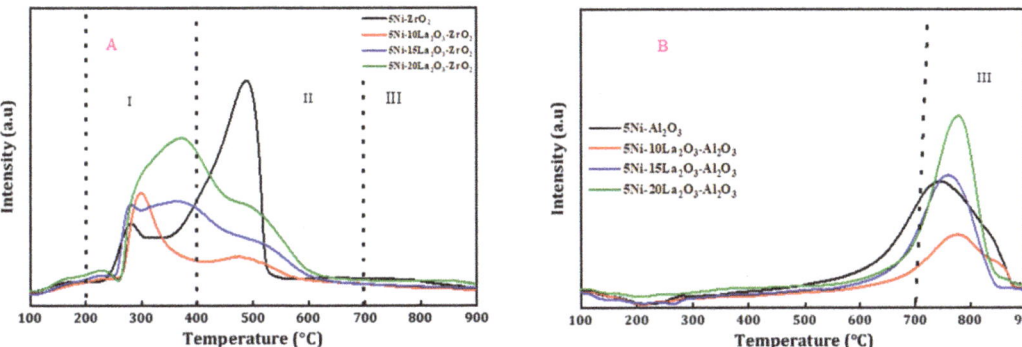

Figure 2. H_2-TPR profiles of (**A**) for 5Ni-xLa$_2$O$_3$-ZrO$_2$ and (**B**) 5Ni-xLa$_2$O$_3$-Al$_2$O$_3$ (x = 0, 10, 15, and 20 wt%) catalysts.

Figure 3 presents the powder X-ray diffraction patterns of the calcined 5Ni-xLa$_2$O$_3$-ZrO$_2$ and 5Ni-xLa$_2$O$_3$-Al$_2$O$_3$ catalysts (x = 0, 10, 15, and 20 wt%). Figure 3A displays the XRD patterns for 5Ni-xLa$_2$O$_3$-ZrO$_2$. There are no peaks attributable to La$_2$O$_3$ in these patterns since similar patterns are obtained for La$_2$O$_3$ modified and un-modified catalysts denoting the homogenous distribution of La$_2$O$_3$. The peaks at 28.3° and 31.6° are attributed to NiO phase (JCPDS No. 47–1049). The ZrO$_2$ support in Figure 3A has two crystalline phases, tetragonal zirconia (t-ZrO$_2$) and monoclinic zirconia (m-ZrO$_2$). The peaks at 25.5° and 34.3° are ascribed to m-ZrO$_2$ [44,45], while the peaks at 40.9°, 50.2°, and 55.5° are credited to t-ZrO$_2$ [46]. In Figure 3B, there is no peak ascribable to La$_2$O$_3$ in the patterns and therefore La$_2$O$_3$ is well dispersed in the alumina matrix. The characteristics peaks at 37.3° (311), 45.6° (400), and 67.0° (440), all are correspondingly allocated to the Al$_2$O$_3$ structure (JCPDS 10–0425) [47,48].

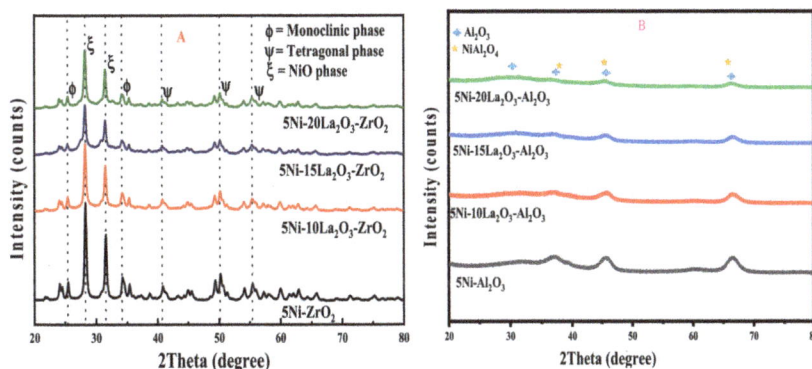

Figure 3. XRD patterns of the (**A**) 5Ni-xLa$_2$O$_3$+ZrO$_2$ (**B**) 5Ni-xLa$_2$O$_3$+Al$_2$O$_3$ catalysts (x = 0, 10, 15, and 20 wt%) calcined at 700 °C.

Figure 4 displays the hydrogen yield versus time on stream for the dry reforming reaction at 700 °C. The impact of La$_2$O$_3$ addition on the DRM catalytic performance of Ni-ZrO$_2$ is discussed in this section. The initial hydrogen yield of the 5Ni-ZrO$_2$ catalyst in Figure 4 is lower than the La$_2$O$_3$ modified catalysts. An evident trend is noted when La$_2$O$_3$ is added causing a significant improvement in the DRM performance. The improvement profile escalates as the following: 5Ni-ZrO$_2$ < 5Ni-20La$_2$O$_3$-ZrO$_2$ < 5Ni-15La$_2$O$_3$-ZrO$_2$ <

5Ni-10La$_2$O$_3$-ZrO$_2$. The highest hydrogen yield of about 80% is recorded using 10% La$_2$O$_3$. The improvement due to the La$_2$O$_3$ addition is attributed to the fact that La$_2$O$_3$ increases the dispersion of Ni particles on the supports and reduces the agglomeration of Ni particles during the reforming reaction as depicted in Figure 2A. Moreover, La$_2$O$_3$ increases the basicity as shown in the TPD profiles and therefore adsorb and react with CO$_2$ to form La$_2$O$_2$CO$_3$ species on the surface of catalyst which can speed up the conversion [49].

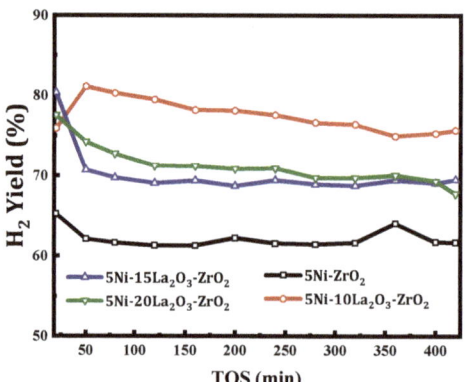

Figure 4. H$_2$ yield vs. time on stream over the 5Ni-xLa$_2$O$_3$-ZrO$_2$ (x = 0, 10, 15, and 20 wt%) catalysts at 700 °C for 420 min.

In Figure 5, the hydrogen yield obtained is close for La$_2$O$_3$ doped and non-doped catalysts. The La$_2$O$_3$ addition improved marginally the hydrogen yield in the following manner 5Ni-Al$_2$O$_3$ < 5Ni-20La$_2$O$_3$-Al$_2$O$_3$ < 5Ni-10La$_2$O$_3$-Al$_2$O$_3$ < 5Ni-15La$_2$O$_3$-Al$_2$O$_3$. The 15% La$_2$O$_3$ gave the highest hydrogen yield of 84%. It can be inferred that the effect of La$_2$O$_3$ loading affects differently the hydrogen yield productivity depending on the type of the support. Table 1 describes the efficiency of the present work and some of the literature.

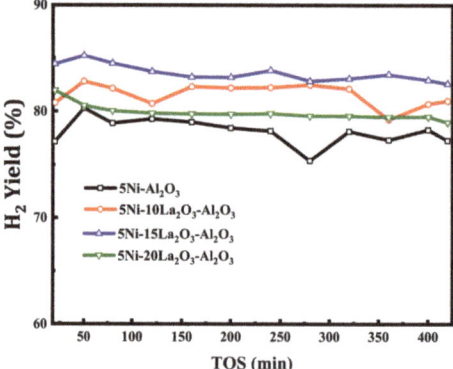

Figure 5. H$_2$ yield vs. time on stream over the 5Ni-xLa$_2$O$_3$-Al$_2$O$_3$ (x = 0, 10, 15, and 20 wt%) catalysts at 700 °C for 420 min.

Table 1. Hydrogen yield performances obtained in CO$_2$ reforming of methane of present and past work.

Catalyst	Feed (CH$_4$:CO$_2$: Inert)	Reaction Temperature (°C)	GHSV (mL/(g·h)	Yield (%)	Ref.
Ni/Zr-γAl$_2$O$_3$	1:1:3	750	54,000	63	[50]
Ni@ZrO$_2$-SiZr-7.7	1:1:1	800	72,000	81	[51]
0.8% Ni+0.2% Co-MgAl$_2$O$_4$	1:1:0.5	700	54,000	51	[52]
10Ni+1%Fe-MgAl$_2$O$_4$	1:1:1	750	30,000	78	[53]
5Ni-10La$_2$O$_3$-ZrO$_2$	1:1:0.33	700	42,000	80	This work
5Ni-15La$_2$O$_3$-Al$_2$O$_3$	1:1:0.33	700	42,000	84	This work

Figure 6 exhibits the CO$_2$-TPD profiles of the (A) 5Ni-xLa$_2$O$_3$-ZrO$_2$ and (B) 5Ni-xLa$_2$O$_3$-Al$_2$O$_3$ (x = 0, 10, 15, and 20 wt%) spent catalysts obtained at 700 °C reaction temperature. This was performed to scan the surface basicity of the catalysts, which plays a vital role in the catalytic DRM reaction [54]. It is commonly understood that a greater desorption temperature of CO$_2$ reveals a stronger basicity and a bigger amount of CO$_2$ desorption although signifying that more basic sites are presented on the surface of catalyst [55]. In Figure 6A, three chief CO$_2$ desorption peaks were identified in the experimented temperature range from 50 to 700 °C for higher loadings of La$_2$O$_3$ (5Ni-15La$_2$O$_3$-ZrO$_2$ and 5Ni-20La$_2$O$_3$-ZrO$_2$) modified catalysts and only two peaks for the un-modified (5Ni-ZrO$_2$) and lower loading La$_2$O$_3$ (5Ni-10La$_2$O$_3$-ZrO$_2$) modified catalysts, which denoted that three types of basic sites existed in the 5Ni-xLa$_2$O$_3$ + ZrO$_2$ (x = 0, 10, 15, and 20 wt%) catalysts. The CO$_2$ desorption peaks appeared at 80 °C, 260 °C and 550 °C. The peaks correspond to the weak adsorption of CO$_2$ on OH groups, moderate adsorption of CO$_2$ and strong CO$_2$ adsorption on the metal–oxygen pairs and O^{2-} anions, respectively [56,57]. It is clear that CO$_2$ was favorably absorbed on the strong basic sites as the support was modified with La$_2$O$_3$, rather than bare zirconia support. This result showed that the adsorption of CO$_2$ had altered from physical adsorption to chemical adsorption because of the addition of La$_2$O$_3$. Figure 6B shows two main CO$_2$ desorption peaks at 80 °C and 250 °C corresponding to weak and moderate basic sites. Hence, the addition of La$_2$O$_3$ to the support did give significant variation in basicity of the 5Ni-xLa$_2$O$_3$ + Al$_2$O$_3$ catalysts. The 5Ni-15La$_2$O$_3$-Al$_2$O$_3$ displays larger peak intensity than the remaining catalysts, which is in accordance with the better activity observed. Table 2 shows a summary of a quantitative assessment of CO$_2$ adsorption of the spent catalysts via CO$_2$-TPD for ZrO$_2$ and Al$_2$O$_3$ supported catalysts.

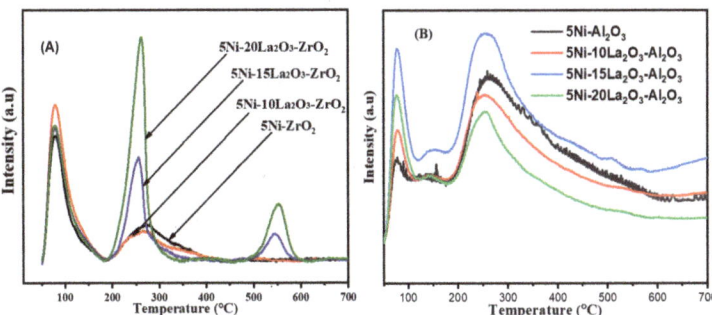

Figure 6. CO$_2$-TPD profiles of the spent (A) 5Ni-xLa$_2$O$_3$-ZrO$_2$ and (B) 5Ni-xLa$_2$O$_3$-Al$_2$O$_3$ (x = 0, 10, 15, and 20 wt%) catalysts obtained at 700 °C reaction temperature.

Table 2. Quantitative assessment of CO$_2$ adsorption of the spent catalysts via CO$_2$-TPD.

Catalyst	Weak-Basicity (μmol/g)	Medium-Basicity (μmol/g)	Strong-Basicity (μmol/g)	Total-Basicity (μmol/g)
5Ni-ZrO$_2$ [a]	0.52	0.48	0.00	1
5Ni-10La$_2$O$_3$-ZrO$_2$	0.66	0.29	0.00	0.95
5Ni-15La$_2$O$_3$-ZrO$_2$	0.63	0.49	0.12	1.24
5Ni-20La$_2$O$_3$-ZrO$_2$	0.52	0.94	0.27	1.73
5Ni-Al$_2$O$_3$ [a]	0.44	0.38	0.18	1
5Ni-10La$_2$O$_3$-Al$_2$O$_3$	0.42	0.11	0.00	0.53
5Ni-15La$_2$O$_3$-Al$_2$O$_3$	0.32	0.10	0.00	0.42
5Ni-20La$_2$O$_3$-Al$_2$O$_3$	0.35	0.06	0.00	0.41

[a] For comparison and basicity assessment of catalysts, the sum of all basic sites of 5Ni-ZrO$_2$ is set to 1 and 5Ni-Al$_2$O$_3$ is set to 1.

Figure 7A,B contain images obtained from Energy-dispersive X-ray spectroscopy (EDX) analysis of the fresh samples of both 5Ni-10La$_2$O$_3$-ZrO$_2$ and 5Ni-15La$_2$O$_3$-Al$_2$O$_3$. These results show the elemental composition of the as-prepared catalyst samples. First, the analysis confirmed the presence of all the elemental constituents that were mixed together during the catalyst synthesis. Moreover, the percentage loadings revealed by the EDX analysis are virtually the same as intended in the calculation and catalyst preparation with error values of 10% for Ni and 6% for La.

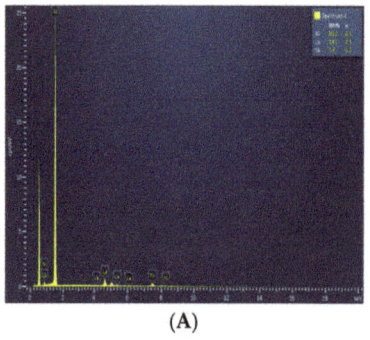 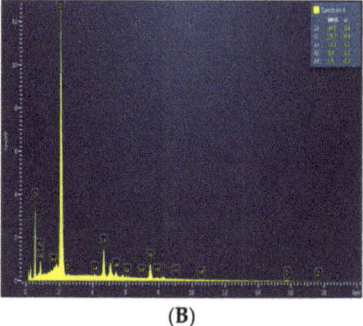

(A) (B)

Figure 7. EDX analysis of fresh (**A**) 5Ni-15La$_2$O$_3$-Al$_2$O$_3$ and (**B**) 5Ni-10La$_2$O$_3$-ZrO$_2$ showing the elemental composition of prepared catalysts.

Broad examination of the morphology of catalysts was performed via TEM. Typical TEM overviews of the fresh catalysts (Figure 8A,C) and spent catalyst samples (Figure 8B,D) were acquired after DRM at 700 °C for 420 min time on stream. There is a difference in the morphology of the fresh and used catalysts. The fresh catalyst seems to have particles clumped on the surface, while the spent catalysts depict rough and dispersed particles on the catalyst surface. In addition to the recognized metal particles, carbon in the form of nanotubes can be seen in the images of the used catalysts. The image of the fresh 5Ni-10La$_2$O$_3$-ZrO$_2$ catalyst is shown in Figure 8A. It is evident from the TEM images that the Ni is homogeneously dispersed over the surface of the support. In contrast to its spent sample as shown in Figure 8B, big quantities of coke are formed commonly in the form of nanotubes. The TEM images showed well dispersed Ni species ((average particle diameter of 4 nm) over La$_2$O$_3$-Al$_2$O$_3$ support (Figure 8C). However, after the reaction (Figure 8D) size of Ni species has grown (average particle diameter of 7–8 nm).

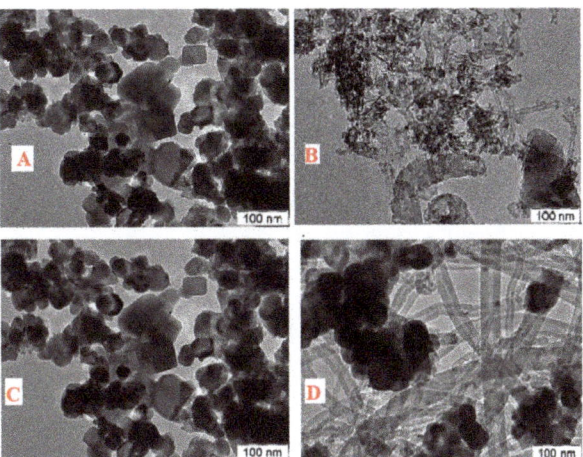

Figure 8. TEM image of (**A**) Fresh, (**B**) spent of 5Ni/10La$_2$O$_3$+ZrO$_2$, (**C**) Fresh, (**D**) spent of 5Ni/15La$_2$O$_3$ + Al$_2$O$_3$ catalysts.

The Fourier Transform Infrared Spectroscopy (FTIR) analysis of fresh 5Ni-ZrO$_2$, 5Ni-10La$_2$O$_3$-ZrO$_2$, 5Ni-Al$_2$O$_3$ and 5Ni-15La$_2$O$_3$-Al$_2$O$_3$ catalysts were done to investigate the existing bonds within the catalyst system. The infrared spectra of absorption for these samples are shown in Figure 9A,B.

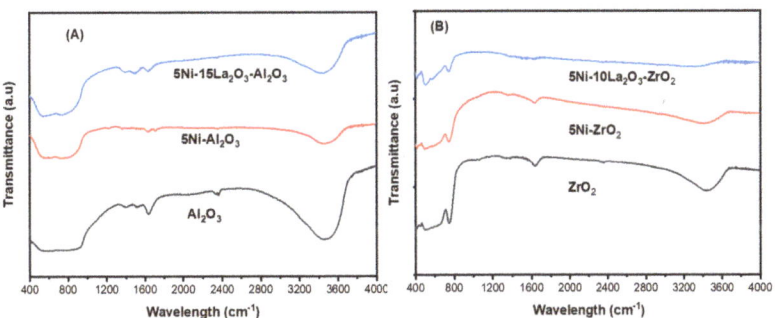

Figure 9. The FTIR spectra of fresh (**A**) Al$_2$O$_3$, 5Ni-Al$_2$O$_3$, 5Ni-15La$_2$O$_3$-Al$_2$O$_3$ and (**B**) ZrO$_2$, 5Ni-ZrO$_2$, 5Ni-10La$_2$O$_3$-ZrO$_2$ showing the existing stretching vibration.

In Figure 9A, Al$_2$O$_3$ is well known for its tendency to adsorb moisture from the atmosphere onto itself. Thus, the distinct band representing the stretching vibration of O-H, within the wavelengths 3430–3460 cm^{-1} for all the samples can be ascribed to [OH]$^{-1}$ groups interaction and/or physisorbed moisture interaction that is adsorbed onto the Al$_2$O$_3$ support [58]. It is noticeable from the figure that the extent of hydration of the catalyst surface changed with metal addition as the OH band's intensity decreased on adding active metal. Moreover, the vibration bands centered at around 1400, 1520, 1649, and 2366 cm^{-1} can be seen for the support and the other samples. This implies that these bands can be associated with the support, the Al–O bond stretching to be specific [59]. The small less noticeable peaks appearing at wavelengths 1237 and 1719 cm^{-1} for the 5Ni-Al$_2$O$_3$ sample can be said to be the stretching vibration of NiO and NiAl$_2$O$_4$ species, respectively. The latter is thought to have stronger interaction with the support, in light of that it appeared at a higher wavelength.

As for the ZrO_2 support and ZrO_2 supported catalysts (Figure 9B); the band within 3357–3440 cm^{-1} can be said to be hydrogen-bonded bending and stretching of the OH groups as a result of adsorbed moisture while the peaks at 1632–1640 cm^{-1} can be assigned to the vibration of the water molecules [60]. These peaks are seen to decrease in intensity with the addition of nickel and lanthanum oxide to the support. The peaks that are centered at around 457 and 752 cm^{-1} represent the asymmetric stretching of the Zr–O–Zr bond [61]. The vibration owing to the presence of La_2O_3 was not discovered. This further supports the results obtained from the XRD analysis of samples with La_2O_3.

Figure 10A displays TGA profiles of spent 5Ni-xLa_2O_3 + ZrO_2 (x = 0, 10, 15, and 20 wt%) catalysts operated at 700 °C. The weight loss above 500 °C was due to the removal of deposited carbon. The extents of carbon deposition on the spent catalysts exhibit the following sequence: 5Ni-10La_2O_3-ZrO_2 < 5Ni-15La_2O_3-ZrO_2 < 5Ni-20La_2O_3-ZrO_2 < 5Ni-ZrO_2. The un-promoted ZrO_2 catalyst gives the highest weight loss of 47.4%. The results are well established with the catalytic performance. Similarly, Figure 9B shows the TGA profiles of spent 5Ni-xLa_2O_3 + Al_2O_3 (x = 0, 10, 15, and 20 wt%) catalysts at 700 °C. The amounts of weight loss are close to each between the La_2O_3 promoted and non-promoted catalysts. Since values range between 9.0% for 5Ni-10La_2O_3-Al_2O_3 and 17.2% for 5Ni-ZrO_2, this indicates the Al_2O_3 supported catalysts are more resistant to carbon deposition than ZrO_2 supported catalysts.

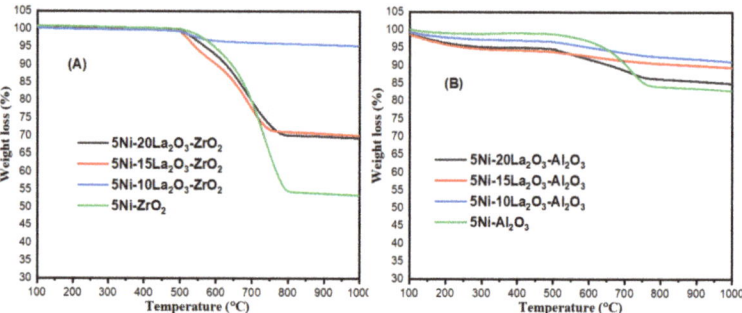

Figure 10. TGA profiles of spent (**A**) 5Ni-xLa_2O_3-ZrO_2, (**B**) 5Ni-xLa_2O_3-Al_2O_3 (x = 0, 10, 15, and 20 wt%) catalysts at 700 °C.

Figure 11 shows the Raman analysis of the used catalysts (5Ni-ZrO_2, 5Ni-10La_2O_3-ZrO_2, 5Ni-Al_2O_3, and 5Ni-15La_2O_3-Al_2O_3). The D (deformation) and G (graphitic) bands appear at nearly 1342 and 1580 cm^{-1} respectively except for spent 5Ni-Al_2O_3 with D and G bands appearing at about 1468 and 1532 cm^{-1}, respectively. The spent catalysts are characterized by carbon deposits of different degree of graphitization. It is established that carbon deposits having high I_G to I_D ratio show better extent of graphitization [62]. From the figure, it can be seen that 5Ni-ZrO_2 had the highest degree of graphitization followed by 5Ni-Al_2O_3. Moreover, it can be inferred that the graphitization decreases with the addition of La_2O_3. Thus, La_2O_3 promotes the formation carbons that are defective.

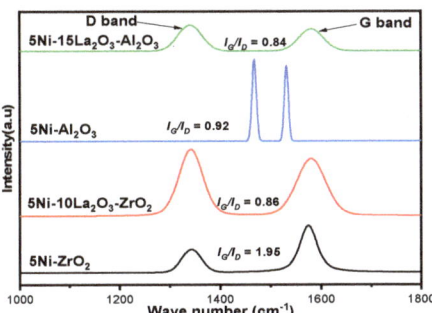

Figure 11. Raman spectra for the spent catalysts (5Ni-ZrO$_2$, 5Ni-10La$_2$O$_3$-ZrO$_2$, 5Ni-Al$_2$O$_3$, and 5Ni-15La$_2$O$_3$-Al$_2$O$_3$) showing the extent of graphitization of the carbon deposits.

4. Conclusions

This paper elucidates the role of La$_2$O$_3$ addition to two different supports (ZrO$_2$ and Al$_2$O$_3$) of Ni-based catalysts for the hydrogen production via CO$_2$ reforming of methane. The 5Ni-10La$_2$O$_3$-ZrO$_2$ catalyst increased the hydrogen yield by 25% in comparison with the pristine 5Ni-ZrO$_2$. Similarly, the 5Ni-15La$_2$O$_3$-Al$_2$O$_3$ catalyst showed improved efficiency of 8% of hydrogen yield. The La$_2$O$_3$ loading influenced differently the ZrO$_2$ and the Al$_2$O$_3$ supports. The study displayed that the modified La$_2$O$_3$-Al$_2$O$_3$ support catalysts gave a higher hydrogen yield than La$_2$O$_3$-ZrO$_2$ supported catalyst. The catalyst characterizations showed that La$_2$O$_3$ addition improved specific surface areas, dispersion, reducibility, metal–support interaction, and surface basic sites which contributed towards the enhanced hydrogen yield. The qualitative and quantitative analysis of carbon formed over the spent catalysts using TEM, TGA, and Raman spectroscopy showed presence of carbon nanotubes. This work provides an insight towards the role of support modification during DRM.

Supplementary Materials: The following are available online at https://www.mdpi.com/article/10.3390/en14092412/s1, Table S1: Textural properties of different catalysts supported Ni catalysts: BET specific surface area (S$_{BET}$), pore volume (P$_V$), and pore diameter (D$_P$), Table S2: The quantitative analysis of H$_2$ consumption during H$_2$-TPR.

Author Contributions: A.A.-F., A.I. and S.K. synthesized the catalysts, carried out all the experiments and characterization tests, and wrote the manuscript. W.K. and M.S. prepared the catalyst and contributed to proofreading of the manuscript. A.A. and A.F. contributed to the analysis of the data and writing—review of the manuscript. All authors have read and agreed to the published version of the manuscript.

Funding: This research was funded by Deanship of Scientific Research at King Saud University so the authors would like to express their sincere appreciation to the Deanship of Scientific Research at King Saud University for funding this research project (#RG-119).

Institutional Review Board Statement: Not applicable.

Informed Consent Statement: Not applicable.

Data Availability Statement: Not applicable.

Acknowledgments: The authors would like to express their sincere appreciation to the Deanship of Scientific Research at King Saud University for funding this research project (#RG-119).

Conflicts of Interest: The authors declare no conflict of interest.

References

1. Wang, Q.; Chen, X.; Jha, A.N.; Rogers, H. Natural gas from shale formation—The evolution, evidences and challenges of shale gas revolution in United States. *Renew. Sustain. Energy Rev.* **2014**, *30*, 1–28. [CrossRef]

2. Montgomery, C.T.; Smith, M.B. Hydraulic Fracturing: History of an Enduring Technology. *J. Pet. Technol.* **2010**, *62*, 26–40. [CrossRef]
3. Themelis, N.J.; Ulloa, P.A. Methane generation in landfills. *Renew. Energy* **2007**, *32*, 1243–1257. [CrossRef]
4. Izquierdo, U.; Barrio, V.; Requies, J.; Cambra, J.; Güemez, M.; Arias, P. Tri-reforming: A new biogas process for synthesis gas and hydrogen production. *Int. J. Hydrog. Energy* **2013**, *38*, 7623–7631. [CrossRef]
5. Al-Fatesh, A.S.; Kumar, R.; Kasim, S.O.; Ibrahim, A.A.; Fakeeha, A.H.; Abasaeed, A.E.; Alrasheed, R.; Bagabas, A.; Chaudhary, M.L.; Frusteri, F.; et al. The effect of modifier identity on the performance of Ni-based catalyst supported on γ-Al_2O_3 in dry reforming of methane. *Catal. Today* **2020**, *348*, 236–242. [CrossRef]
6. Ibrahim, A.A.; Al-Fatesh, A.S.; Khan, W.U.; Kasim, S.O.; Abasaeed, A.E.; Fakeeha, A.H.; Bonura, G.; Frusteri, F. Enhanced coke suppression by using phosphate-zirconia supported nickel catalysts under dry methane reforming conditions. *Int. J. Hydrog. Energy* **2019**, *44*, 27784–27794. [CrossRef]
7. Al-Fatesh, A.S.; Abu-Dahrieh, J.K.; Atia, H.; Armbruster, U.; Ibrahim, A.A.; Khan, W.U.; Abasaeed, A.E.; Fakeeha, A.H. Effect of pre-treatment and calcination temperature on Al_2O_3-ZrO_2 supported Ni-Co catalysts for dry reforming of methane. *Int. J. Hydrog. Energy* **2019**, *44*, 21546–21558. [CrossRef]
8. Zhang, S.; Muratsugu, S.; Ishiguro, N.; Tada, M. Ceria-Doped Ni/SBA-16 Catalysts for Dry Reforming of Methane. *ACS Catal.* **2013**, *3*, 1855–1864. [CrossRef]
9. Boyano, A.; Morosuk, T.; Blanco-Marigorta, A.; Tsatsaronis, G. Conventional and advanced exergoenvironmental analysis of a steam methane reforming reactor for hydrogen production. *J. Clean. Prod.* **2012**, *20*, 152–160. [CrossRef]
10. Kim, A.R.; Lee, H.Y.; Cho, J.M.; Choi, J.-H.; Bae, J.W. Ni/M-Al_2O_3 (M=Sm, Ce or Mg) for combined steam and CO_2 reforming of CH_4 from coke oven gas. *J. CO2 Util.* **2017**, *21*, 211–218. [CrossRef]
11. Lougou, B.G.; Shuai, Y.; Chaffa, G.; Xing, H.; Tan, H.; Du, H. Analysis of CO_2 utilization into synthesis gas based on solar thermochemical CH_4-reforming. *J. Energy Chem.* **2019**, *28*, 61–72. [CrossRef]
12. Wu, Z.; Yang, B.; Miao, S.; Liu, W.; Xie, J.; Lee, S.; Pellin, M.J.; Xiao, D.; Su, D.; Ma, D. Lattice Strained Ni-Co alloy as a High-Performance Catalyst for Catalytic Dry Reforming of Methane. *ACS Catal.* **2019**, *9*, 2693–2700. [CrossRef]
13. Arora, S.; Prasad, R. An overview on dry reforming of methane: Strategies to reduce carbonaceous deactivation of catalysts. *RSC Adv.* **2016**, *6*, 108668–108688. [CrossRef]
14. Varga, E.; Pusztai, P.; Steinrück, H.-P.; Óvári, L.; Papp, C.; Kónya, Z.; Oszkó, A.; Erdőhelyi, A.; Kiss, J. Probing the interaction of Rh, Co and bimetallic Rh–Co nanoparticles with the CeO_2 support: Catalytic materials for alternative energy generation. *Phys. Chem. Chem. Phys.* **2015**, *17*, 27154–27166. [CrossRef]
15. Pakhare, D.; Spivey, J. A review of dry (CO_2) reforming of methane over noble metal catalysts. *Chem. Soc. Rev.* **2014**, *43*, 7813–7837. [CrossRef] [PubMed]
16. Damyanova, S.; Shtereva, I.; Pawelec, B.; Mihaylov, L.; Fierro, J.L.G. Characterization of none and yttrium-modified Ni-based catalysts for dry reforming of methane. *Appl. Catal. B Environ.* **2020**, *278*, 119335. [CrossRef]
17. Jalali, R.; Rezaei, M.; Nematollahi, B.; Baghalha, M. Preparation of Ni/$MeAl_2O_4$-$MgAl_2O_4$ (Me=Fe, Co, Ni, Cu, Zn, Mg) nanocatalysts for the syngas production via combined dry reforming and partial oxidation of methane. *Renew. Energy* **2020**, *149*, 1053–1067. [CrossRef]
18. Liu, C.-J.; Ye, J.; Jiang, J.; Pan, Y. Progresses in the Preparation of Coke Resistant Ni-based Catalyst for Steam and CO_2 Reforming of Methane. *ChemCatChem* **2011**, *3*, 529–541. [CrossRef]
19. Wang, N.; Shen, K.; Yu, X.; Qian, W.; Chu, W. Preparation and characterization of a plasma treated NiMgSBA-15 catalyst for methane reforming with CO_2 to produce syngas. *Catal. Sci. Technol.* **2013**, *3*, 2278–2287. [CrossRef]
20. Bradford, M.C.; Vannice, M.A. The role of metal–support interactions in CO_2 reforming of CH_4. *Catal. Today* **1999**, *50*, 87–96. [CrossRef]
21. Tang, M.; Liu, K.; Roddick, D.M.; Fan, M. Enhanced lattice oxygen reactivity over Fe_2O_3/Al_2O_3 redox catalyst for chemical-looping dry (CO_2) reforming of CH_4: Synergistic La-Ce effect. *J. Catal.* **2018**, *368*, 38–52. [CrossRef]
22. Faria, E.; Neto, R.; Colman, R.; Noronha, F. Hydrogen production through CO2 reforming of methane over Ni/$CeZrO_2$/Al_2O_3 catalysts. *Catal. Today* **2014**, *228*, 138–144. [CrossRef]
23. Stagg-Williams, S.M.; Noronha, F.B.; Fendley, G.; Resasco, D.E. CO_2 Reforming of CH_4 over Pt/ZrO_2 Catalysts Promoted with La and Ce Oxides. *J. Catal.* **2000**, *194*, 240–249. [CrossRef]
24. Li, Z.-X.; Shi, F.-B.; Li, L.-L.; Zhang, T.; Yan, C.-H. A facile route to ordered mesoporous-alumina-supported catalysts, and their catalytic activities for CO oxidation. *Phys. Chem. Chem. Phys.* **2010**, *13*, 2488–2491. [CrossRef] [PubMed]
25. Dehimi, L.; Benguerba, Y.; Virginie, M.; Hijazi, H. Microkinetic modelling of methane dry reforming over Ni/Al_2O_3 catalyst. *Int. J. Hydrog. Energy* **2017**, *42*, 18930–18940. [CrossRef]
26. Lašič Jurković, D.; Liu, J.-L.; Pohar, A.; Likozar, B. Methane Dry Reforming over Ni/Al_2O_3 Catalyst in Spark Plasma Reactor: Linking Computational Fluid Dynamics (CFD) with Reaction Kinetic Modelling. *Catal. Today* **2021**, *362*, 11–21. [CrossRef]
27. Bian, Z.; Zhong, W.; Yu, Y.; Wang, Z.; Jiang, B.; Kawi, S. Dry reforming of methane on Ni/mesoporous-Al_2O_3 catalysts: Effect of calcination temperature. *Int. J. Hydrog. Energy* **2021**. [CrossRef]
28. Newnham, J.; Mantri, K.; Amin, M.H.; Tardio, J.; Bhargava, S.K. Highly stable and active Ni-mesoporous alumina catalysts for dry reforming of methane. *Int. J. Hydrog. Energy* **2012**, *37*, 1454–1464. [CrossRef]

29. Montoya, J.; Romero-Pascual, E.; Gimon, C.; Del Angel, P.; Monzón, A. Methane reforming with CO_2 over Ni/ZrO_2–CeO_2 catalysts prepared by sol–gel. *Catal. Today* **2000**, *63*, 71–85. [CrossRef]
30. Li, W.; Zhao, Z.; Jiao, Y. Dry reforming of methane towards CO-rich hydrogen production over robust supported Ni catalyst on hierarchically structured monoclinic zirconia nanosheets. *Int. J. Hydrog. Energy* **2016**, *41*, 17907–17921. [CrossRef]
31. Jang, J.T.; Yoon, K.J.; Bae, J.W.; Han, G.Y. Cyclic production of syngas and hydrogen through methane-reforming and water-splitting by using ceria–zirconia solid solutions in a solar volumetric receiver–reactor. *Sol. Energy* **2014**, *109*, 70–81. [CrossRef]
32. Hu, X.; Jia, X.; Zhang, X.; Liu, Y.; Liu, C.-J. Improvement in the activity of Ni/ZrO_2 by cold plasma decomposition for dry reforming of methane. *Catal. Commun.* **2019**, *128*, 105720. [CrossRef]
33. Zhang, M.; Zhang, J.; Zhou, Z.; Chen, S.; Zhang, T.; Song, F.; Zhang, Q.; Tsubaki, N.; Tan, Y.; Han, Y. Effects of the surface adsorbed oxygen species tuned by rare-earth metal doping on dry reforming of methane over Ni/ZrO_2 catalyst. *Appl. Catal. B Environ.* **2020**, *264*, 118522. [CrossRef]
34. Zhang, W.; Liu, B.; Zhu, C.; Tian, Y. Preparation of La_2NiO_4/ZSM-5 catalyst and catalytic performance in CO_2/CH_4 reforming to syngas. *Appl. Catal. A Gen.* **2005**, *292*, 138–143. [CrossRef]
35. Horváth, A.; Stefler, G.; Geszti, O.; Kienneman, A.; Pietraszek, A.; Guczi, L. Methane dry reforming with CO_2 on CeZr-oxide supported Ni, NiRh and NiCo catalysts prepared by sol–gel technique: Relationship between activity and coke formation. *Catal. Today* **2011**, *169*, 102–111. [CrossRef]
36. Tran, N.T.; Van Le, Q.; Van Cuong, N.; Nguyen, T.D.; Phuc, N.H.H.; Phuong, P.T.; Monir, M.U.; Aziz, A.A.; Truong, Q.D.; Abidin, S.Z.; et al. La-doped cobalt supported on mesoporous alumina catalysts for improved methane dry reforming and coke mitigation. *J. Energy Inst.* **2020**, *93*, 1571–1580. [CrossRef]
37. Liu, H.; Hadjltaief, H.B.; Benzina, M.; Gálvez, M.E.; Da Costa, P. Natural clay based nickel catalysts for dry reforming of methane: On the effect of support promotion (La, Al, Mn). *Int. J. Hydrog. Energy* **2019**, *44*, 246–255. [CrossRef]
38. Yabe, T.; Mitarai, K.; Oshima, K.; Ogo, S.; Sekine, Y. Low-temperature dry reforming of methane to produce syngas in an electric field over La-doped Ni/ZrO_2 catalysts. *Fuel Process. Technol.* **2017**, *158*, 96–103. [CrossRef]
39. Al-Fatesh, A.S.; Arafat, Y.; Ibrahim, A.A.; Kasim, S.O.; Alharthi, A.; Fakeeha, A.H.; Abasaeed, A.E.; Bonura, G.; Frusteri, F. Catalytic Behaviour of Ce-Doped Ni Systems Supported on Stabilized Zirconia under Dry Reforming Conditions. *Catalysts* **2019**, *9*, 473. [CrossRef]
40. Mandal, S.; Santra, C.; Kumar, R.; Pramanik, M.; Rahman, S.; Bhaumik, A.; Maity, S.; Sen, D.; Chowdhury, B. Niobium doped hexagonal mesoporous silica (HMS-X) catalyst for vapor phase Beckmann rearrangement reaction. *RSC Adv.* **2014**, *4*, 845–854. [CrossRef]
41. Rahman, S.; Santra, C.; Kumar, R.; Bahadur, J.; Sultana, A.; Schweins, R.; Sen, D.; Maity, S.; Mazumdar, S.; Chowdhury, B. Highly active Ga promoted Co-HMS-X catalyst towards styrene epoxidation reaction using molecular O_2. *Appl. Catal. A Gen.* **2014**, *482*, 61–68. [CrossRef]
42. Jiang, X.-Y.; Zhou, R.-X.; Pan, P.; Zhu, B.; Yuan, X.-X.; Zheng, X.-M. Effect of the addition of La_2O_3 on TPR and TPD of $CuO\gamma$-Al_2O_3 catalysts. *Appl. Catal. A Gen.* **1997**, *150*, 131–141. [CrossRef]
43. Mo, W.; Ma, F.; Ma, Y.; Fan, X. The optimization of Ni–Al_2O_3 catalyst with the addition of La_2O_3 for CO_2–CH_4 reforming to produce syngas. *Int. J. Hydrog. Energy* **2019**, *44*, 24510–24524. [CrossRef]
44. Goula, M.; Charisiou, N.; Siakavelas, G.; Tzounis, L.; Tsiaoussis, I.; Panagiotopoulou, P.; Goula, G.; Yentekakis, I. Syngas production via the biogas dry reforming reaction over Ni supported on zirconia modified with CeO_2 or La_2O_3 catalysts. *Int. J. Hydrog. Energy* **2017**, *42*, 13724–13740. [CrossRef]
45. Zhang, J.; Gao, Y.; Jia, X.; Wang, J.; Chen, Z.; Xu, Y. Oxygen vacancy-rich mesoporous ZrO_2 with remarkably enhanced visible-light photocatalytic performance. *Sol. Energy Mater. Sol. Cells* **2018**, *182*, 113–120. [CrossRef]
46. Titus, J.; Roussière, T.; Wasserschaff, G.; Schunk, S.; Milanov, A.; Schwab, E.; Wagner, G.; Oeckler, O.; Gläser, R. Dry reforming of methane with carbon dioxide over NiO–MgO–ZrO_2. *Catal. Today* **2016**, *270*, 68–75. [CrossRef]
47. Alipour, Z.; Rezaei, M.; Meshkani, F. Effects of support modifiers on the catalytic performance of Ni/Al_2O_3 catalyst in CO_2 reforming of methane. *Fuel* **2014**, *129*, 197–203. [CrossRef]
48. Mierczynski, P.; Mierczynska, A.; Ciesielski, R.; Mosinska, M.; Nowosielska, M.; Czylkowska, A.; Maniukiewicz, W.; Szynkowska, M.I.; Vasilev, K. High Active and Selective Ni/CeO_2–Al_2O_3 and Pd–Ni/CeO_2–Al_2O_3 Catalysts for Oxy-Steam Reforming of Methanol. *Catalysts* **2018**, *8*, 380. [CrossRef]
49. Li, K.; Pei, C.; Li, X.; Chen, S.; Zhang, X.; Liu, R.; Gong, J. Dry reforming of methane over $La_2O_2CO_3$-modified Ni/Al_2O_3 catalysts with moderate metal support interaction. *Appl. Catal. B Environ.* **2020**, *264*, 118448. [CrossRef]
50. Khani, Y.; Bahadoran, F.; Shariatinia, Z.; Varmazyari, M.; Safari, N. Synthesis of highly efficient and stable $Ni/Ce_xZr_{1-x}Gd_xO_4$ and Ni/X-Al_2O_3 (X = Ce, Zr, Gd, Ce-Zr-Gd) nanocatalysts applied in methane reforming reactions. *Ceram. Int.* **2020**, *46*, 25122–25135. [CrossRef]
51. Lim, Z.-Y.; Tu, J.; Xu, Y.; Chen, B. $Ni@ZrO_2$ yolk-shell catalyst for CO_2 methane reforming: Effect of $Ni@SiO_2$ size as the hard-template. *J. Colloid Interface Sci.* **2021**, *590*, 641–651. [CrossRef]
52. Aghaali, M.H.; Firoozi, S. Enhancing the catalytic performance of Co substituted $NiAl_2O_4$ spinel by ultrasonic spray pyrolysis method for steam and dry reforming of methane. *Int. J. Hydrog. Energy* **2021**, *46*, 357–373. [CrossRef]

53. Medeiros, R.L.; Macedo, H.P.; Melo, V.R.; Oliveira, Â.A.S.; Barros, J.M.; Melo, M.A.; Melo, D.M. Ni supported on Fe-doped MgAl$_2$O$_4$ for dry reforming of methane: Use of factorial design to optimize H$_2$ yield. *Int. J. Hydrog. Energy* **2016**, *41*, 14047–14057. [CrossRef]
54. Lavoie, J.-M. Review on dry reforming of methane, a potentially more environmentally-friendly approach to the increasing natural gas exploitation. *Front. Chem.* **2014**, *2*, 81. [CrossRef]
55. Zhang, L.; Zhang, Q.; Liu, Y.; Zhang, Y. Dry reforming of methane over Ni/MgO-Al$_2$O$_3$ catalysts prepared by two-step hydrothermal method. *Appl. Surf. Sci.* **2016**, *389*, 25–33. [CrossRef]
56. Wang, H.; Zhao, B.; Qin, L.; Wang, Y.; Yu, F.; Han, J. Non-thermal plasma-enhanced dry reforming of methane and CO$_2$ over Ce-promoted Ni/C catalysts. *Mol. Catal.* **2020**, *485*, 110821. [CrossRef]
57. Zheng, X.; Li, Y.; Zhang, L.; Shen, L.; Xiao, Y.; Zhang, Y.; Au, C.; Jiang, L. Insight into the effect of morphology on catalytic performance of porous CeO$_2$ nanocrystals for H$_2$S selective oxidation. *Appl. Catal. B Environ.* **2019**, *252*, 98–110. [CrossRef]
58. Lei, H.; Zhang, P. Preparation of alumina/silica core-shell abrasives and their CMP behavior. *Appl. Surf. Sci.* **2007**, *253*, 8754–8761. [CrossRef]
59. Sanchez, E.A.; Comelli, R.A. Hydrogen by glycerol steam reforming on a nickel–alumina catalyst: Deactivation processes and regeneration. *Int. J. Hydrog. Energy* **2012**, *37*, 14740–14746. [CrossRef]
60. Goharshadi, E.K.; Hadadian, M. Effect of calcination temperature on structural, vibrational, optical, and rheological properties of zirconia nanoparticles. *Ceram. Int.* **2012**, *38*, 1771–1777. [CrossRef]
61. Heshmatpour, F.; Aghakhanpour, R.B. Synthesis and characterization of superfine pure tetragonal nanocrystalline sulfated zirconia powder by a non-alkoxide sol–gel route. *Adv. Powder Technol.* **2012**, *23*, 80–87. [CrossRef]
62. Kim, S.; Crandall, B.S.; Lance, M.J.; Cordonnier, N.; Lauterbach, J.; Sasmaz, E. Activity and stability of NiCe@SiO multi–yolk–shell nanotube catalyst for tri-reforming of methane. *Appl. Catal. B Environ.* **2019**, *259*, 118037. [CrossRef]

Article

Syngas Production via CO₂ Reforming of Methane over SrNiO₃ and CeNiO₃ Perovskites

Naushad Ahmad [1,*], Fahad Alharthi [1], Manawwer Alam [1], Rizwan Wahab [2], Salim Manoharadas [3] and Basel Alrayes [4]

1. Department of Chemistry, College of Science, King Saud University, P.O. Box 2454, Riyadh 11451, Saudi Arabia; fharthi@ksu.edu.sa (F.A.); maalam@ksu.edu.sa (M.A.)
2. Department of Zoology, College of Science, King Saud University, P.O. Box 2454, Riyadh 11451, Saudi Arabia; rwahab@ksu.edu.sa
3. Central Laboratory, Department of Botany and Microbiology, College of Science, King Saud University, P.O. Box 2454, Riyadh 11451, Saudi Arabia; smanoharadas@ksu.edu.sa
4. Central Laboratory, College of Science, King Saud University, P.O. Box 2454, Riyadh 11451, Saudi Arabia; bfalrayes@ksu.edu.sa
* Correspondence: anaushad@ksu.edu.sa or naushaddrnaima@gmail.com

Abstract: The development of a transition-metal-based catalyst with concomitant high activity and stability due to its distinguishing characteristics, yielding an abundance of active sites, is considered to be the bottleneck for the dry reforming of methane (DRM). This work presents the catalytic activity and durability of $SrNiO_3$ and $CeNiO_3$ perovskites for syngas production via DRM. $CeNiO_3$ exhibits a higher specific surface area, pore volume, number of reducible species, and nickel dispersion when compared to $SrNiO_3$. The catalytic activity results demonstrate higher CH_4 (54.3%) and CO_2 (64.8%) conversions for $CeNiO_3$, compared to 22% (CH_4 conversion) and 34.7% (CO_2 conversion) for $SrNiO_3$. The decrease in catalytic activity after replacing cerium with strontium is attributed to a decrease in specific surface area and pore volume, and nickel active sites covered with strontium carbonate. The stability results reveal the deactivation of both the catalysts ($SrNiO_3$ and $CeNiO_3$) but $SrNiO_3$ showed more deactivation than $CeNiO_3$, as demonstrated by deactivation factors. The catalyst deactivation is mainly attributed to carbon deposition and these findings are verified by characterizing the spent catalysts.

Keywords: perovskites; strontium; cerium; hydrogen; sintering; carbon deposition

1. Introduction

Dry, or carbon dioxide reforming of methane (DRM) has gained attention in recent decades, mainly due to the fact that DRM consumes prevalent greenhouse gases i.e., methane and carbon dioxide to produce synthetic gas, which serves as an important raw material for liquid hydrocarbon formation [1–8]. Hence, DRM offers two benefits: (a) conversion of major greenhouse gases into a value-added product, and (b) the DRM product, i.e., syngas, offers equimolar H_2 and CO, which results in hydrocarbon production via Fischer–Tropsch (FT) synthesis [9–13]. The catalytic activity and stability are mainly dependent on the choice of a suitable catalyst [14]. Non-noble-metal-based catalysts, particularly transition-metal-based catalysts including catalysts of Ni, and Co, are mostly studied for DRM since these catalysts offer advantages such as their abundance, quick turnover rates, and low cost [15–18]. The bottlenecks associated with Ni-based catalysts include the loss of active metal surface area due to sintering and carbon formation during DRM which results in catalyst deactivation and also influences the selectivity of the syngas produced [19].

Generally, the basic supports or promoters, such as CeO_2, La_2O_3, and Sr^{2+}, have demonstrated better catalytic activity and enhanced chemisorption of CO_2 than acidic supports. Many researchers have reported that ceria and its modified supported catalysts provide a promising platform for endothermic DRM processes due to their basicity, to promote CO_2 adsorption, and their high oxygen storage capacity/oxygen vacancy for CO_2 activation or the gasification of different kinds of carbon precursors [20–23]. Perovskites have shown excellent performance in catalytic and photovoltaic industries and Ni-based perovskites are favored for DRM as perovskites offer high metal dispersion and thermal stability [24,25]. The perovskites in which the B-site cation is replaced with transition metals such as Ni need to be researched in depth [26,27]. Generally, several factors contribute to the catalytic performance of a perovskite [28], (a) the choice of element(s) for B-site cation, (b) controlling vacancy and valency through the proper selection of A-site element(s) and/or partially substituting companion metal(s), (c) high dispersion obtained due to the formation of fine particles, which leads to higher specific surface area, (d) the synergy between A-site and B-site elements.

Ren et al. [29] investigated the role of an Mo_2C-Ni/ZrO_2 catalyst in the steam–CO_2 dual reformation of methane and found the catalyst exhibited high catalytic activity (~75% CH_4 conversion) and unexpected coke-resistant stability, as evidenced by TGA, even after 30 h time-on-stream. In other research, $LaBO_3$ (B = Ni, Fe, Co, and Mn) perovskites were studied for the reduction in pollution from vehicles fueled with natural gas [30]. Moreover, the effect of adding Pd to $LaBO_3$ perovskites on oxidation activity performance showed that a smaller amount of Pd contributes to improving not only lattice oxygen mobility, but also enhances the reducibility of the B-site in $LaBPd_{0.05}O_3$ perovskites. Hence, Pd addition significantly enhanced catalytic activity of the perovskites.

Messaoudi et al. [31] studied the role of bulk La_xNiO_y and supported La_xNiO_y/$MgAl_2O_4$ catalysts in DRM and found that the supported catalysts exhibited higher nickel dispersions and specific surface areas. These factors contributed to enhanced activity and stability, with minimal carbon formation during a 65 h time-on-stream. They also discovered that the supported catalysts had intact metallic nickel active sites after a long-term stability test, as verified by XRD results. The study of the impact of catalyst preparation methods, gas hourly space velocity, and reaction temperatures on catalytic performance of ternary perovskites $AZrRuO_3$ (A = Ca, Ba, and Sr) revealed that the $SrZrRuO_3$ catalyst exhibited the highest conversion and best stability among the tested perovskites [32]. Wang et al. [33] utilized perovskite (La_2O_3-$LaFeO_3$) as the support to load Ni and Co to synthesize bimetallic catalysts, to explore their performance in DRM. The loading of a suitable amount of Co increased the catalytic activity and suppressed carbon deposition, which is attributed to the crystalline structure of the perovskite.

In this work, $SrNiO_3$ and $CeNiO_3$ perovskites were synthesized and investigated for CO_2 reformation of methane. The activity performance in terms of CH_4 and CO_2 conversions and the catalyst durability, i.e., activity as function of time, are specifically elucidated herein. Overall, the aim of the study is to provide insights into the replacement of cerium with strontium and their respective responses to perovskite activity and stability under reforming conditions. The catalysts were characterized before and after activity and stability tests to understand and discuss the catalytic findings in relation to analysis results.

2. Materials and Methods

2.1. Preparation of $SrNiO_3$ and $CeNiO_3$ Nanocrystals

Nanocrystals of $SrNiO_3$ and $CeNiO_3$ were synthesized by the self-combustion method using metallic nitrates and glycine as a precursor. Firstly, 1 mmol $Ce(NO_3)_3 \cdot nH_2O$ (Sigma Aldrich, St. Louis, MO, USA—99.9%), 1 mmol $Ni(NO_3)_2 \cdot 6H_2O$ (Sigma Aldrich, St. Louis, MO, USA—99.9%), and 1 mmol $Sr(NO_3)_2$ were separately dissolved in 100 mL deionized water. The solutions of Ce/Sr and Ni were mixed in a 1:1 ratio to obtain a clear and homogeneous solution. Then, glycine (purity 99.5%), used as an ignition promoter, was added to the metal nitrate solutions (glycine:metal ions ≈ 1). The mixtures were thoroughly stirred by a magnetic mixer to eliminate the water at 60–70 °C until a homogeneous sol-like solution was formed. The gel was heated up to around 250 °C, at which temperature the ignition reaction occurred, producing a powdered precursor which still contained some carbon residue. Finally, the powders were calcined in air at 700 °C for 6 h to eliminate the remained carbon, resulting in the formation of the perovskite structure.

2.2. Catalyst Analysis

The thermogravimetric curves (TG-DTG) of the dried precursors, from ambient temperature to 1000 °C under nitrogen flow (100 mL min^{-1}) at a heating rate of 20 °C min^{-1}, were recorded on Mettler-Toledo TGA/SDTA851e thermal analyzer, (Schwerzenbach Switzerland). All techniques mentioned below were employed on calcined powders. X-ray reflection patterns from 10–80° at a scan rate of 0.2° min^{-1} were recorded on a Shimadzu XRD-6000 diffractometer (Columbia, MD, USA) with monochromatic radiation of CuK (λ = 1.5406 Å). The specific surface area was measured by nitrogen adsorption on a Quantachrome NOVA 2000e BET system (Boynton Beach, FL, USA) and the pore size was measured by the BJH method. Temperature programmed reduction (TPR) experiments were carried out on a semiautomatic Micromeritics 2920 apparatus (Norcross, GA, USA). Samples of about 30 mg were placed in a U-shaped quartz tube, first purged in a synthetic air stream of 50 mL min^{-1} at 300 °C for 1 h and then cooled to ambient temperature. Reduction profiles were then recorded by passing a 10% H$_2$/Air flow over the samples at a rate of 25 mL min^{-1}, while heating at a rate of 10 °C min^{-1} from ambient temperature to 900 °C. Temperature programmed oxidation (TPO) was performed on the catalysts after the dry reforming stability tests, using the same instruments with which the TPR was performed, to verify the carbon formation. Transmission electron microscopy (TEM) was carried out using a JEOL JEM-1011 microscope (Tokyo, Japan) with an accelerating voltage of 80 kV. The samples were then placed in a copper grid where the liquid phase was evaporated. They were then used to analyze the morphology of the fresh and used catalysts for the estimation of deposited carbon.

2.3. Catalyst Activity Measurements

The dry reforming of methane was carried out in a tubular fixed-bed stainless steel reactor (i.d. 10 mm, catalytic bed length 3 mm) coupled to a gas chromatograph (Shimadzu HP5890 series II, Kyoto, Japan) with a thermal conductivity detector. The reaction conditions were: 700 °C, 0.05 g of catalyst, a gas mixture CH$_4$:CO$_2$:N$_2$ (30:30:10, 70 mL/min), a space velocity of 84,000 mLg$_{cat.}$$^{-1}$h^{-1}, and at 1 atm pressure. Prior to the catalytic activity tests, the catalysts were reduced in a 10% H$_2$/N$_2$ mixture (40 mL/min) at 700 °C for 120 min. Then, the H$_2$ flow was replaced by a He flow (60 mL/min), and the system was heated (10 °C/min) to the reaction temperature. Stability runs were carried out at 700 °C for periods of 440 min time-on-stream. The products of reactions were analyzed on-line by a VARIAN GC 3800 gas-chromatograph (Varian, Santa Clara, CA, USA), equipped with two thermal conductivity detectors and columns packed with Porapak N and 13X Molecular sieves (Varian, Santa Clara, CA, USA). The reproducibility of the gas phase composition was checked in replica experiments. In most experiments, the error was within 5%.

3. Results and Discussion

3.1. Characterization of As-Synthesized Catalysts

3.1.1. Thermal Decomposition of Precursors

Figure 1 shows the TG and DTG thermal decomposition curves of the precursors which are used to determine the final calcination temperature for the formation of crystalline products. Figure 1 shows that during the decomposition process, many phases are formed, but overall CeNiO$_3$ exhibits a two step, and SrNiO$_3$ a three step, decomposition. For CeNiO$_3$, the weight loss (~8%) started slowly at about ~75 °C, reached a maximum rate at ~150 °C (T1) and was finally completed at ~300 °C. The weight loss below this temperature was caused by the removal of the water left over from crystallization and the release of gases. The weight loss (~21%) from 300 to 550 °C with a maximum rate at 420 °C (T2 in DTG curve), may be regarded as a result of the decomposition and burning of the remaining organic matter. Further heating caused negligible weight loss with the release of minute gaseous products in the form of CO_2 and formation of the perovskite phase.

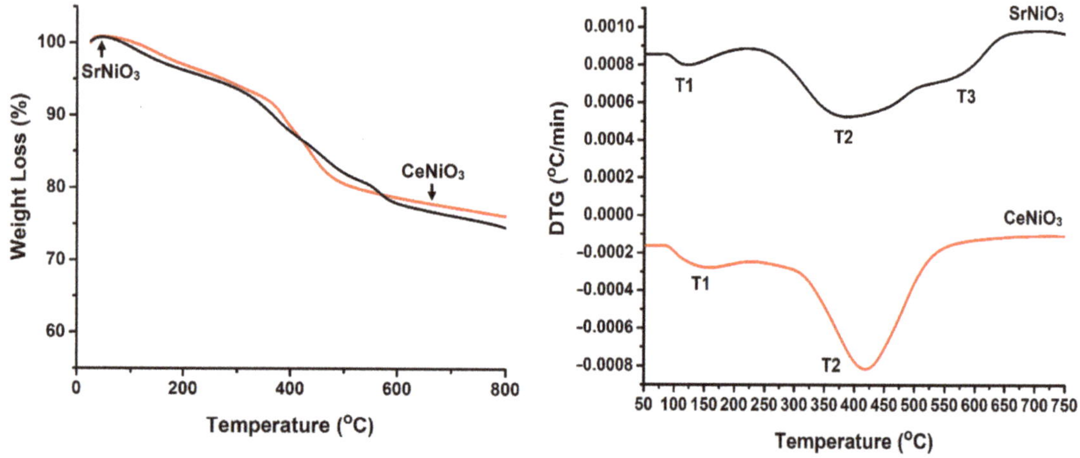

Figure 1. TG-DTG curves versus temperature of $SrNiO_3$ and $CeNiO_3$ perovskites.

On the contrary, for $SrNiO_3$, three step decomposition was observed at different intervals of temperature. The initial weight loss below 125 °C was attributed to the loss of water and some adsorbed gases. The other two steps of decomposition were attributed to the combustion of organic matter present in the precursor. Therefore, from the TG-DTG curves of the fabricated samples, it can be inferred that the perovskite phase forms above 700 °C.

3.1.2. X-ray Diffraction (XRD)

The XRD profiles of the as-synthesized $SrNiO_3$ and $CeNiO_3$ perovskites are presented in Figure 2. Upon analyzing diffraction data using MDI Jade® software (version 6.5, Materials data Inc., Newtown Square, PA, USA), it was found that diffraction peaks corresponding to 2θ values of 28.6°, 33.2°, 37.3°, 43.3°, 47.5°, 56.4°, 62.9°, 69.5°, and 76.7° are assigned to crystal planes (111), (200), (111), (200), (220), (111), and (110) of $CeNiO_3$, respectively [34,35]. The peaks appearing at 2θ = 28.6°, 33.2°, 47.5°, and 56.4° are related to cubic CeO_2 corresponding to crystal planes (111), (200), (220), and (311), respectively (JCPDS 81–0792). The peaks observed for cubic NiO were found to be at 2θ = 37.3°, 43.3°, and 62.9°, corresponding to crystal planes (111), (200), and (220), respectively (JCPDS 75–0197). Additionally, peaks ascribed to SrO and $SrCO_3$, as labelled in Figure 2, are also observed for $SrNiO_3$ perovskite [36].

3.1.3. Textural Properties

Table 1 shows the BET surface areas and pore parameters for $CeNiO_3$ and $SrNiO_3$ perovskites. The observed values of 20.7 m²/g ($CeNiO_3$) and 12.2 m²/g ($SrNiO_3$) are rather low and similar to those previously reported for perovskite-type oxides calcined at 800 °C, which is higher than is generally reported in the literature (below 10 m²/g) for this type of material [37]. It is observed that the surface area of $CeNiO_3$ catalyst is higher (20.7 m²/g) than $SrNiO_3$ catalyst (12.2 m²/g). A similar trend is also observed for pore volume. Hence, $CeNiO_3$ is expected to show higher activity than $SrNiO_3$.

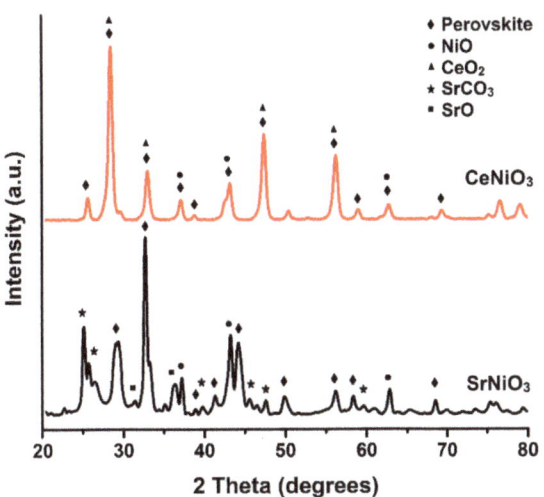

Figure 2. XRD patterns of SrNiO$_3$ and CeNiO$_3$ perovskites.

Table 1. DTG decomposition temperatures of SrNiO$_3$ and CeNiO$_3$ perovskites.

Perovskite	T1 (°C)	T2 (°C)	T3 (°C)
CeNiO$_3$	150	420	-
SrNiO$_3$	125	390	570

3.1.4. Morphological Study (TEM) of Fresh Perovskites

The morphology of as-synthesized SrNiO$_3$ and CeNiO$_3$ perovskites was analyzed by using Transmission Electron Microscopy (TEM), and microscopic images of all the fresh, reduced, and used catalysts are shown in Figure 3. The analysis of TEM images was performed using ImageJ® software (version 1.5, National Institutes of Health, Bethesda, MD, USA). The spherical particles were found to have average sizes varying from 5 to 34 nm, and 13 to 24 nm for fresh CeNiO$_3$, and SrNiO$_3$ perovskites (Figure 3a,d), respectively, which increased slightly to 8–45 nm and 16–39 nm for CeNiO$_3$, and SrNiO$_3$ perovskites (Figure 3b,e), respectively, after reduction, indicating negligible sintering after activation.

3.1.5. Temperature-Programmed Reduction (TPR)

Temperature-programmed reduction or TPR is a handy tool to analyze the reducibility–metal-support interaction and to find the reduction or activation temperature required to generate metallic particles prior to catalytic reaction. The reduction profiles shown in Figure 4 indicate the variation in reducibility and metal–support interaction when Ce is replaced with Sr. The small reduction peak, below 150 °C for the CeNiO$_3$ catalyst is related to the reduction in adsorbed oxygen species. Reduction peaks appearing between 200 and 500 °C correspond to the reduction of Ni^{3+} to Ni^{2+}, while the small shoulders at higher temperatures (>500 °C) are attributed to the reduction of Ni^{2+} to Ni^0 [38,39]. On the contrary, significant changes in the reduction peak temperatures (~245 and 345 °C) were observed for SrNiO$_3$. Additionally, the three-fold decrease in peak height of SrNiO$_3$, when compared to CeNiO$_3$, indicates that number of reducible species was suppressed by the replacement of Ce with Sr. This was also evident from the total amounts of hydrogen consumed during TPR (Table 2). The degree of reduction was significantly lower in SrNiO$_3$ (50.6%) than CeNiO$_3$ (Table 2), which can be attributed to poor dispersion of Ni within the SrNiO$_3$. The reduction in both CeNiO$_3$ and SrNiO$_3$ can be expressed as (CeNiO$_3$ + H$_2$ → Ni^0 + CeO$_2$ + H$_2$O) and (SrNiO$_3$ + 2H$_2$ → Ni^0 + SrO + 2H$_2$O). The role of these findings in influencing the catalytic activity is discussed in Section 3.

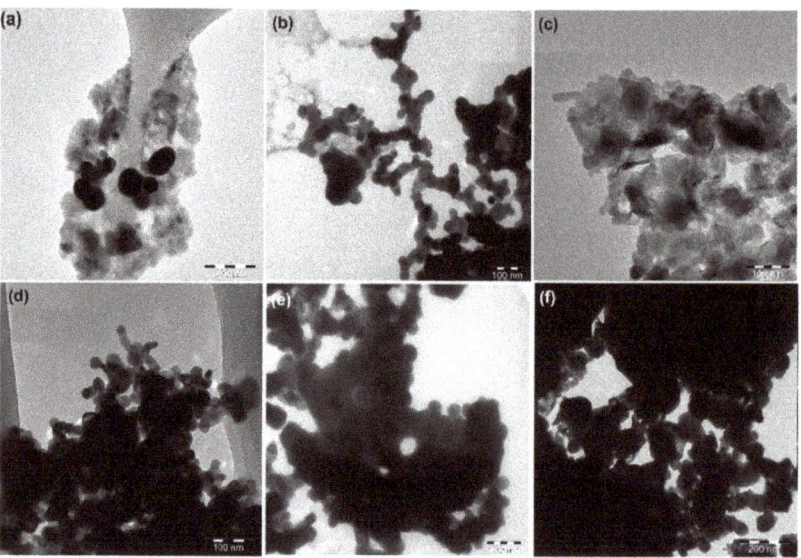

Figure 3. TEM images of (**a**) fresh CeNiO$_3$, (**b**) reduced CeNiO$_3$, (**c**) used CeNiO$_3$, (**d**) fresh SrNiO$_3$, (**e**) reduced SrNiO$_3$, and (**f**) used SrNiO$_3$.

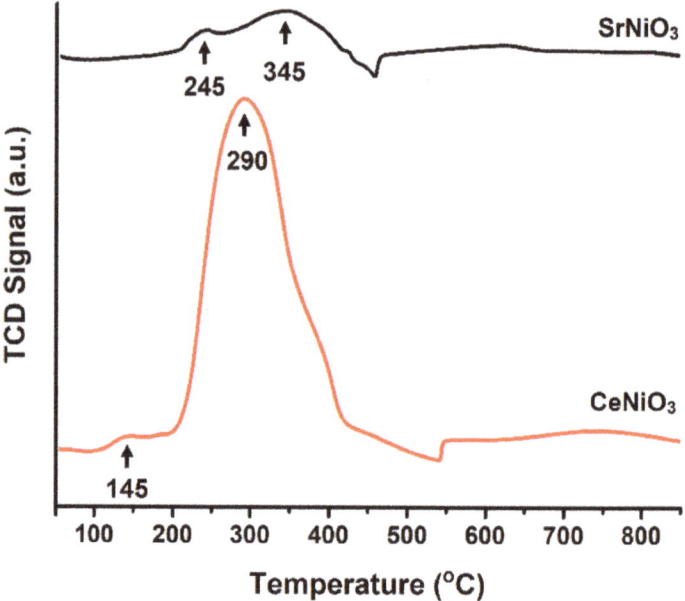

Figure 4. H$_2$-TPR profiles of SrNiO$_3$ and CeNiO$_3$ perovskites.

Table 2. Textural properties and deactivation factors of SrNiO$_3$ and CeNiO$_3$ perovskites.

Perovskite	S_{BET} (m^2/g)	Pore Volume (cm^3/g)	Pore Size (nm)	Deactivation Factor (%) [a]	Total Hydrogen Consumption (mmol/g) [b]	Degree of Reduction (%) [c]
CeNiO$_3$	20.7	0.162	30.1	7.7	1.11	90.5
SrNiO$_3$	12.2	0.026	9.3	64.7	0.84	50.6

[a] Deactivation Factor (D.F., %) = 100 × (CH$_4$ conversion$_{initial}$ − CH$_4$ conversion$_{final}$)/(CH$_4$ conversion$_{initial}$); [b] from TPR; [c] the ratio of amount of hydrogen consumed in TPR to the theoretical amount of hydrogen required to completely reduce the catalyst.

3.2. Catalytic Performances

The as-synthesized SrNiO$_3$ and CeNiO$_3$ perovskites were investigated for their catalytic performance at 700 °C. Due to the fact that the dry reforming reaction requires metallic nickel crystallites as active sites, all the perovskites were reduced under a hydrogen atmosphere prior to the reaction study. The activity results, in terms of CH$_4$ and CO$_2$ conversions as a function of time, are shown in Figure 5a,b, respectively. From Figure 5a, it is evident that CeNiO$_3$ deactivates over time, despite displaying relatively higher CH$_4$ conversion. CeNiO$_3$ demonstrates an initial CH$_4$ conversion of 54.3% which reaches 50.1% after 440 min time-on-stream, resulting in deactivation factor of 7.7% (Table 2). Strontium incorporation clearly influences CH$_4$ conversion, as initial conversion decreased from 54.3% (CeNiO$_3$) to 22% (SrNiO$_3$). Interestingly, significant deactivation is observed for the strontium incorporated perovskite (SrNiO$_3$) and hence a deactivation factor of 64.4% is found for SrNiO$_3$ (Table 2). The deactivation of the catalysts and the factors behind it are analyzed by characterizing the spent catalysts, as discussed in Section 3.3. A similar trend was found in Figure 5b for CO$_2$ conversions versus time-on-stream. Initial CO$_2$ conversions of 64.8 and 34.7% were demonstrated by CeNiO$_3$ and SrNiO$_3$, respectively, which reached final conversions of 58 and 11.5%, respectively. It is also worth observing from Figure 5 that CO$_2$ conversions are higher than those of CH$_4$. This result implies the simultaneous existence of the reverse water-gas shift reaction (CO$_2$ + H$_2$ → CO + H$_2$O) that generates H$_2$/CO molar ratios lower than the stoichiometric one (H$_2$/CO = 1.0) due to the fact that hydrogen consumes CO$_2$ and CO in a disproportionation or Boudouard reaction (2CO → CO$_2$ + C). A separate section is dedicated to the discussion of catalytic activity results in relation to their analysis findings in Section 3.4.

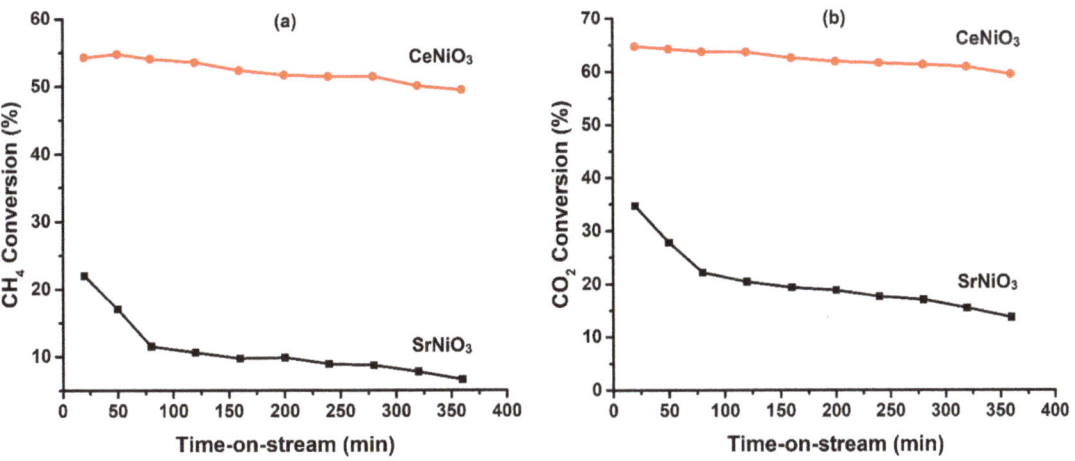

Figure 5. (a) CH$_4$ conversion, and (b) CO$_2$ conversion versus time-on-stream (TOS) of SrNiO$_3$ and CeNiO$_3$ perovskites.

3.3. Characterization of Spent Perovskites

The perovskites, after being investigated for dry reforming reaction, were further analyzed to understand their catalytic performance results. $CeNiO_3$ showed deactivation while all the perovskites showed CO_2 conversions higher than CH_4 conversions (Figure 5), which gave rise to side reactions such as reverse water-gas shift and CO disproportionation. The perovskites were then analyzed using temperature-programmed oxidation and transition electron microscope to assess the modifications to perovskites during reaction.

3.3.1. Temperature-Programmed Oxidation (TPO)

In order to verify the possibility of carbon deposition over the surface of the perovskites, TPO analysis was carried out after the reforming reaction. Figure 6 presents the TPO results of the $SrNiO_3$ and $CeNiO_3$ perovskites. Both of the perovskites showed one broad peak in the temperature range of from 170 to 550 °C. The peak maximum temperatures are ~320 and 335 °C for $SrNiO_3$ and $CeNiO_3$, respectively. These peaks were attributed to the polymeric species of carbon deposited and/or less reactive surface carbides formed during the reaction, as reported earlier [40,41]. The peak temperatures correspond to the degree of hydrogenation of surface carbon species and the surface carbon changing to be graphitic in nature as the peak temperature increased. It is evident from peak temperatures that both perovskites have shown the formation of mainly polymeric carbon species and that their interaction with the catalyst surface changes, becoming stronger with strontium replacement by cerium, as demonstrated by the increase in peak temperatures from 320 to 335 °C. It could also be observed that cerium oxide played a role in controlling the carbon deposition, carbon alleviation and the degree of interaction between carbon and the catalyst surface.

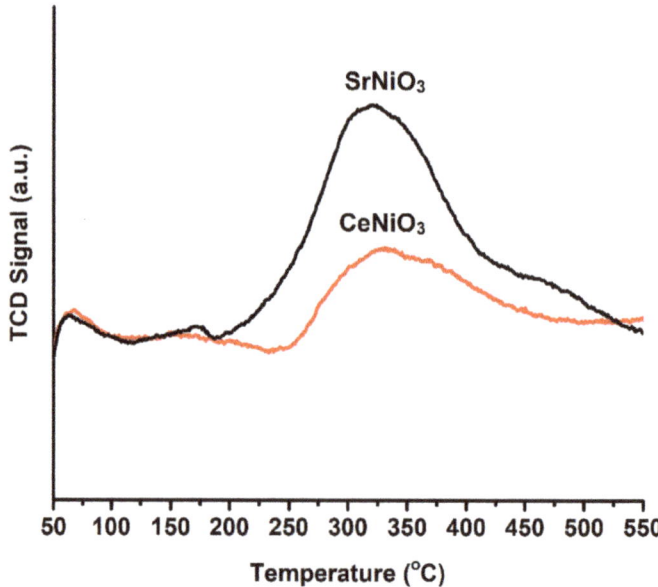

Figure 6. TPO profiles of $SNiO_3$ and $CeNiO_3$ perovskites.

3.3.2. Transition Electron Microscopy (TEM)

To further verify the causes of deactivation or modification of the perovskites during the reforming reaction, transition electron microscopic analysis was carried out. The TEM images in Figure 3c,f show particle sizes of 10–70 nm and 25–77 nm for $CeNiO_3$ and $SrNiO_3$ spent catalysts, respectively. The TEM results manifest the formation of carbon over spent catalysts as well as noticeable agglomeration of the perovskite particles. Hence, sintering also contributes to the deactivation of $SrNiO_3$ and $CeNiO_3$. These findings are discussed below in Section 3.4.

3.4. Discussion

The as-synthesized perovskites were analyzed prior to reaction to predict their performance during reaction. TG-DTG data show that the precursors were converted into perovskite after being calcined at the temperatures demonstrated by the DTGs (Figure 1 and Table 1). Both $SrNiO_3$ and $CeNiO_3$ perovskites were formed when they were calcined at 700 °C. XRD diffraction patterns (Figure 2) have shown the existence of perovskite structure and oxides of nickel and cerium along with oxide and carbides of strontium. The textural properties (Table 2), analyzed using nitrogen adsorption-desorption isotherms, demonstrated specific surface areas of 12.2 and 20.7 m^2/g ($SrNiO_3$ and $CeNiO_3$, respectively). Morphological analysis using TEM (Figure 3) displayed spherical particles with different sizes as strontium is replaced with cerium. The TPR profiles (Figure 4) aimed to find out the reduction behavior of the perovskites and it was evident that the reduction in oxides of nickel was easier for $CeNiO_3$, while it became difficult in the case of $SrNiO_3$ which is in agreement with the TG-DTG results (Figure 1). Based on the analyses of perovskites prior to the reforming reaction, it was inferred that $CeNiO_3$ exhibited a higher specific surface area, number of reducible species, and a wider range of particle sizes in comparison with $SrNiO_3$. Hence, CeNiO3 perovskite was expected to show higher activity which was evidenced by CH_4 and CO_2 conversions (Figure 5). It is well known that the dry reforming reaction mechanism needs adsorption of reactants on the active sites, which then dissociate and react to give products, followed by product desorption [4]. Metallic nickel is the main active site for CH_4 adsorption. From catalyst activity results, it was found that $SrNiO_3$ showed a decrease in both CH_4 and CO_2 conversions which can be attributed to the loss of nickel active sites due to agglomeration during calcination, and/or the covering of nickel with strontium oxide or carbonate. From the TPR and TEM images, it can be seen that $SrNiO_3$ has higher reduction temperatures and there is evidence of sintering during calcination. Similarly, the XRD patterns show that clear peaks of oxides and carbonates of strontium are found for $SrNiO_3$ perovskite, which supports the hypothesis that nickel active sites are covered.

Figure 5 also shows the activity results as a function of reaction time which is associated with the durability of the perovskites. It is evident that both perovskites showed deactivation over time, which can be attributed to both sintering and carbon deposition, as evidenced by the TPO and TEM images of the spent catalysts. The extent of the sintering is almost the same for both perovskites during DRM. This suggests that the carbon deposition resulted from methane decomposition, which is a prevalent side reaction at high reaction temperatures and is considered to be the main cause of deactivation. This is in agreement with the TPO results, where it is evident that carbon gasification over the surface of $CeNiO_3$ requires a higher temperature when compared to $SrNiO_3$. Moreover, CO_2 conversions for $CeNiO_3$ are higher than for $SrNiO_3$ which implies that the oxidative environment suitable for carbon gasification is predominant in $CeNiO_3$, suggesting easier carbon removal and no deactivation in this catalyst. Rynkowski et al. [42] investigated DRM in reduced $La_{2-x}Sr_xNiO_4$ perovskite oxides and concluded that the smaller amounts of strontium exhibited less activity and more stability when compared to strontium-free catalysts. Choudhary et al. [43] also studied the influence of the partial substitution of La and Ni in $LaNiO_3$ perovskites and found that catalytic activity is lost after La is partially substituted by Sr in $LaNiO_3$ perovskite.

4. Conclusions

This study investigated the activity and stability performance of $SrNiO_3$ and $CeNiO_3$ perovskites for DRM. The analysis results of $CeNiO_3$ prior to the reaction revealed well-dispersed nickel nanoparticles over the catalyst's surface, enhanced number of reducible species, and higher specific surface areas and pore volumes, which remained key factors in influencing both catalytic activity and durability. The $CeNiO_3$ perovskite demonstrated higher CH_4 and CO_2 conversions as compared to $SrNiO_3$ but both perovskites deactivated over time. Lower activity in the case of $SrNiO_3$ reveals the fact that nickel active sites are covered with strontium carbonates, which is in agreement with previously reported results. The analyses of the perovskites after reaction assisted in locating the cause of deactivation. Though all perovskites showed significant sintering, it was not considered to be the main cause of deactivation as $SrNiO_3$ showed more deactivation despite a similar extent of sintering. Hence, carbon deposition, as evidenced by the TEM and TPO images of spent perovskites, was the main deactivation factor. The investigation of recyclability, reactivation of the developed catalyst system, and the role of strontium combined with Ce-based perovskites are planned for future work.

Author Contributions: N.A., F.A., and M.A. synthesized the catalysts, carried out all the experiments and characterization tests, and wrote the manuscript. R.W., B.A., and S.M. prepared the catalyst and contributed to proofreading of the manuscript. N.A. and F.A. contributed to the analysis of the data and the writing and review of the manuscript. All authors have read and agreed to the published version of the manuscript.

Funding: This research was funded by King Saud University under NPST project (14-PET851-02).

Institutional Review Board Statement: Not Applicable.

Informed Consent Statement: Not Applicable.

Data Availability Statement: Not Applicable.

Acknowledgments: The authors would like to extend their sincere appreciation to the King Saud University for its funding to this NPST project (14-PET851-02).

Conflicts of Interest: The authors declare no conflict of interest.

References

1. Wang, Y.; Yao, L.; Wang, S.; Mao, D.; Hu, C. Low-temperature catalytic CO_2 dry reforming of methane on Ni-based catalysts: A review. *Fuel Process. Technol.* **2018**, *169*, 199–206. [CrossRef]
2. Li, H.; He, Y.; Shen, D.; Cheng, S.; Wang, J.; Liu, H.; Xing, C.; Shan, S.; Lu, C.; Yang, R. Design an in-situ reduction of $Ni/C-SiO_2$ catalyst and new insight into pretreatment effect for $CH_4–CO_2$ reforming reaction. *Int. J. Hydrogen Energy* **2017**, *42*, 10844–10853. [CrossRef]
3. Al-Fatesh, A.S.; Arafat, Y.; Atia, H.; Ibrahim, A.A.; Ha, Q.L.M.; Schneider, M.; M-Pohl, M.; Fakeeha, A.H. CO_2-reforming of methane to produce syngas over Co-Ni/SBA-15 catalyst: Effect of support modifiers (Mg, La and Sc) on catalytic stability. *J. CO2 Util.* **2017**, *21*, 395–404. [CrossRef]
4. Ibrahim, A.A.; Fakeeha, A.H.; Al-Fatesh, A.S. Enhancing hydrogen production by dry reforming process with strontium promoter. *Int. J. Hydrogen Energy* **2014**, *39*, 1680–1687. [CrossRef]
5. Li, D.; Lu, M.; Xu, S.; Chen, C.; Zhan, Y.; Jiang, L. Preparation of supported Co catalysts from Co-Mg-Al layered double hydroxides for carbon dioxide reforming of methane. *Int. J. Hydrogen Energy* **2017**, *42*, 5063–5071. [CrossRef]
6. Al-Fatesh, A.S.; Arafat, Y.; Ibrahim, A.A.; Atia, H.; Fakeeha, A.H.; Armbruster, U.; Abasaeed, A.E.; Frusteri, F. Evaluation of Co-Ni/Sc-SBA–15 as a novel coke resistant catalyst for syngas production via CO_2 reforming of methane. *Appl. Catal. A Gen.* **2018**, *567*, 102–111. [CrossRef]
7. Djinovic, P.; Batista, J.; Pintar, A. Efficient catalytic abatement of greenhouse gases: Methane reforming with CO_2 using a novel and thermally stable $Rh-CeO_2$ catalyst. *Int. J. Hydrogen Energy* **2012**, *37*, 2699–2707. [CrossRef]
8. El Hassan, N.; Kaydouh, M.N.; Geagea, H.; El Zein, H.; Jabbour, K.; Casale, S.; El Zakhem, H.; Massiani, P. Low temperature dry reforming of methane on rhodium and cobalt based catalysts: Active phase stabilization by confinement in mesoporous SBA-15. *Appl. Catal. A Gen.* **2016**, *520*, 114–121. [CrossRef]
9. Khan, W.U.; Li, X.; Baharudin, L.; Yip, A.C.K. Copper-promoted cobalt/titania nanorod catalyst for CO hydrogenation to hydrocarbons. *Catal. Lett.* **2021**, in press. [CrossRef]
10. Khan, W.U.; Baharudin, L.; Choi, J.; Yip, A.C.K. Recent progress in CO hydrogenation over bimetallic catalysts for higher alcohol synthesis. *ChemCatChem* **2020**, *13*, 533–542. [CrossRef]
11. Khan, W.U.; Chen, S.S.; Tsang, D.C.W.; Hu, X.; Lam, F.L.Y.; Yip, A.C.K. Catalytically active interfaces in titania nanorod-supported copper catalysts for CO oxidation. *Nano Res.* **2021**, *13*, 111–120. [CrossRef]
12. Horn, R.; Schlögl, R. Methane activation by heterogeneous catalysis. *Catal. Lett.* **2015**, *45*, 23–39. [CrossRef]

13. Rostrup-Nielsen, J.R.; Sehested, J.; Nørskov, J.K. Hydrogen and synthesis gas by steam- and CO_2 reforming. *Adv. Catal.* **2002**, *47*, 65–139. [CrossRef]
14. Kang, D.; Yu, J.; Ma, W.; Zheng, M.; He, Y.; Li, P. Synthesis of Cu/Ni-$La_{0.7}Sr_{0.3}Cr_{0.5}Mn_{0.5}O_{3-\delta}$ and its catalytic performance on dry methane reforming. *J. Rare Earth* **2019**, *37*, 585–593. [CrossRef]
15. Chein, R.Y.; Fung, W.Y. Syngas production via dry reforming of methane over CeO_2 modified Ni/Al_2O_3 catalysts. *Int. J. Hydrogen Energy* **2019**, *44*, 14303–14315. [CrossRef]
16. Ma, Q.; Guo, L.; Fang, Y.; Li, H.; Zhang, J.; Zhao, T.S.; Yang, G.; Yoneyam, Y.; Tsubaki, N. Combined methane dry reforming and methane partial oxidation for syngas production over high dispersion Ni based mesoporous catalyst. *Fuel Process. Technol.* **2019**, *188*, 98–104. [CrossRef]
17. Abdullah, B.; Ghani, N.A.A.; Vo, D.V.N. Recent advances in dry reforming of methane over Ni-based catalysts. *J. Clean. Prod.* **2017**, *162*, 170–185. [CrossRef]
18. Ali, S.; Khader, M.M.; Almarri, M.J.; Abdelmoneim, A.G. Ni-based nano-catalysts for the dry reforming of methane. *Catal. Today* **2020**, *343*, 26–37. [CrossRef]
19. Shen, J.; Reule, A.A.C.; Semagina, N. $Ni/MgAl_2O_4$ catalyst for low-temperature oxidative dry methane reforming with CO_2. *Int. J. Hydrogen Energy* **2019**, *44*, 4616–4629. [CrossRef]
20. Wang, F.; Han, K.; Yu, W.; Zhao, L.; Wang, Y.; Wang, X.; Yu, H.; Shi, W. Low temperature CO_2 reforming with methane reaction over CeO_2-modified $Ni@SiO_2$ catalysts. *ACS Appl. Mater. Interf.* **2020**, *12*, 35022–35034. [CrossRef] [PubMed]
21. Han, K.; Yu, W.; Xu, L.; Deng, Z.; Yu, H.; Wang, F. Reducing carbon deposition and enhancing reaction stability by ceria for methane dry reforming over $Ni@SiO_2@CeO_2$ catalyst. *Fuel* **2021**, *291*, 120182. [CrossRef]
22. Charisiou, N.D.; Siakavelas, G.; Papageridis, K.N.; Baklavaridis, A.; Tzounis, L.; Avraam, D.G.; Goula, M.A. Syngas production via the biogas dry reforming reaction over nickel supported on modified with CeO_2 and/or La_2O_3 alumina catalysts. *J. Nat. Gas. Sci. Eng.* **2016**, *31*, 164–183. [CrossRef]
23. Goula, M.A.; Charisiou, N.D.; Siakavelas, G.; Tzounis, L.; Tsiaoussis, I.; Panagiotopoulou, P.; Goula, G.; Yentekakis, I.V. Syngas production via the biogas dry reforming reaction over Ni supported on zirconia modified with CeO_2 or La_2O_3 catalysts. *Int. J. Hydrogen Energy* **2017**, *42*, 13724–13740. [CrossRef]
24. Liu, M.; Zhang, H.; Gedamu, D.; Fourmont, P.; Rekola, H.; Hiltunen, A.; Cloutier, S.G.; Nechache, R.; Priimagi, A.; Vivo, P. Halide perovskite nanocrystals for next-generation optoelectronics. *Small* **2019**, *15*, 1900801. [CrossRef]
25. Zhang, H.; Deng, R.; Wang, J.; Li, X.; Chen, Y.-M.; Liu, K.; Taubert, C.J.; Cheng, S.Z.D.; Zhu, X. Crystalline organic pigment-based field-effect transistors. *ACS Appl. Mater. Interf.* **2017**, *9*, 21891–21899. [CrossRef]
26. Khalesi, A.; Arandiyan, H.R.; Parvari, M. Production of Syngas by CO_2 Reforming on $M_xLa_{1-x}Ni_{0.3}Al_{0.7}O_{3-d}$ (M = Li, Na, K). *Ind. Eng. Chem. Res.* **2008**, *47*, 5892–5898. [CrossRef]
27. Pérez-Camacho, M.N.; Abu-Dahrieh, J.; Goguet, A.; Sun, K.; Rooney, D. Self-cleaning perovskite-type catalysts for the dry reforming of methane. *Chin. J. Catal.* **2014**, *35*, 1337–1346. [CrossRef]
28. Voorhoeve, R.J.H. 5—Perovskite-related oxides as oxidation—Reduction catalysts. In *Advanced Materials in Catalysis*; Burton, J.J., Garten, R.L., Eds.; Academic Press: New York, NY, USA, 1977; pp. 129–180.
29. Ren, P.; Zhao, Z. Unexpected coke-resistant stability in steam-CO_2 dual reforming of methane over the robust Mo_2C-Ni/ZrO_2 catalyst. *Catal. Commun.* **2019**, *119*, 71–75. [CrossRef]
30. Ziaei-Azad, H.; Khodadadi, A.; Esmaeilnejad-Ahranjani, P.; Mortazavi, Y. Effects of Pd on enhancement of oxidation activity of $LaBO_3$ (B = Mn, Fe, Co and Ni) perovskite catalysts for pollution abatement from natural gas fueled vehicles. *Appl. Catal. B Environ.* **2011**, *102*, 62–70. [CrossRef]
31. Messaoudi, H.; Thomas, S.; Djaidja, A.; Slyemi, S.; Barama, A. Study of La_xNiO_y and $La_xNiO_y/MgAl_2O_4$ catalysts in dry reforming of methane. *J. CO2 Util.* **2018**, *24*, 40–49. [CrossRef]
32. Ruocco, C.; Caprariis, B.D.; Palma, V.; Petrullo, A.; Ricca, A.; Scarsella, M.; Filippis, P.D. Methane dry reforming on Ru perovskites, $AZrRuO_3$: Influence of preparation method and substitution of A cation with alkaline earth metals. *J. CO2 Util.* **2019**, *30*, 222–231. [CrossRef]
33. Wang, H.; Dong, X.; Zhao, T.; Yu, H.; Li, M. Dry reforming of methane over bimetallic Ni-Co catalyst prepared from La$(Co_xNi_{1-x})_{0.5}Fe_{0.5}O_3$ perovskite precursor: Catalytic activity and coking resistance. *Appl. Catal. B Environ.* **2009**, *245*, 302–313. [CrossRef]
34. Dehghani, F.; Ayatollahi, S.; Bahadorikhalili, S.; Esmaeilpour, M. Synthesis and characterization of mixed–metal oxide nanoparticles ($CeNiO_3$, $CeZrO_4$, $CeCaO_3$) and application in adsorption and catalytic oxidation–decomposition of asphaltenes with different chemical structures. *Pet. Chem.* **2020**, *60*, 731–743. [CrossRef]
35. Harikrishnan, M.P.; Mary, A.J.C.; Bose, A.C. Electrochemical performance of $ANiO_3$ (A = La, Ce) perovskite oxide material and its device performance for supercapattery application. *Electrochim. Acta* **2020**, *362*, 137095. [CrossRef]
36. García de la Cruz, R.M.; Falcón, H.; Peña, M.A.; Fierro, J.L.G. Role of bulk and surface structures of $La_{1-x}Sr_xNiO_3$ perovskite-type oxides in methane combustion. *Appl. Catal. B Environ.* **2001**, *33*, 45–55. [CrossRef]
37. Wang, Y.; Cui, X.; Li, Y.; Shu, Z.; Chen, H.; Shi, J. A simple co-nanocasting method to synthesize high surface area mesoporous $LaCoO_3$ oxides for CO and NO oxidations. *Microporous Mesoporous Mater.* **2013**, *176*, 8–15. [CrossRef]

38. Wei, Y.; Zhao, Z.; Jiao, J.; Liu, J.; Duan, A.; Jiang, G. Facile synthesis of three-dimensionally ordered macroporous LaFeO$_3$-supported gold nanoparticle catalysts with high catalytic activity and stability for soot combustion. *Catal. Today* **2015**, *245*, 37–45. [CrossRef]
39. Chen, J.; He, Z.; Li, G.; An, T.; Shi, H.; Li, Y. Visible-light-enhanced photothermocatalytic activity of ABO$_3$-type perovskites for the decontamination of gaseous styrene. *Appl. Catal. B Environ.* **2017**, *209*, 146–154. [CrossRef]
40. Moral, A.; Reyero, I.; Alfaro, C.; Bimbela, F.; Gandía, L.M. Syngas production by means of biogas catalytic partial oxidation and dry reforming using Rh-based catalysts. *Catal. Today* **2018**, *299*, 280–288. [CrossRef]
41. Verykios, X. Mechanistic aspects of the reaction of CO$_2$ reforming of methane over Rh/Al$_2$O$_3$ catalyst. *Appl. Catal. A. Gen.* **2003**, *255*, 101–111. [CrossRef]
42. Rynkowski, J.; Samulkiewicz, P.; Ladavos, A.K.; Pomonis, P.J. Catalytic performance of reduced La$_{2-x}$Sr$_x$NiO$_4$ perovskite-like oxides for CO$_2$ reforming of CH$_4$. *Appl. Catal. A Gen.* **2004**, *263*, 1–9. [CrossRef]
43. Choudhary, V.R.; Uphade, B.S.; Belhekar, A.A. Oxidative conversion of methane to syngas over LaNiO$_3$ perovskite with or without simultaneous steam and CO$_2$ reforming reactions: Influence of partial substitution of La and Ni. *J. Catal.* **1996**, *163*, 312–318. [CrossRef]

Article

Synthesis of BiF$_3$ and BiF$_3$-Added Plaster of Paris Composites for Photocatalytic Applications

V. P. Singh [1,2,†], Mirgender Kumar [3,†], Moolchand Sharma [1], Deepika Mishra [4], Kwang-Su Seong [3], Si-Hyun Park [3,*] and Rahul Vaish [1,*]

1 School of Engineering, Indian Institute of Technology Mandi, Suran 175005, HP, India; vinay.phy@gmail.com (V.P.S.); sharma.moolchand09@gmail.com (M.S.)
2 Department of Physics, Government Engineering College, Bharatpur 321001, RJ, India
3 Department of Electronics Engineering, Yeungnam University, Gyeongsan 38541, Korea; mkumar@ynu.ac.kr (M.K.); kssung@ynu.ac.kr (K.-S.S.)
4 Department of Physics, Sri Satya Sai University of Technology & Medical Sciences, Sehore 466001, MP, India; vinay.matsci@ymail.com
* Correspondence: sihyun_park@ynu.ac.kr (S.-H.P.); rahul@iitmandi.ac.in (R.V.)
† First and second authors are equally contributed.

Abstract: A BiF$_3$ powder sample was prepared from the purchased Bi$_2$O$_3$ powder via the precipitation route. The photocatalytic performance of the prepared BiF$_3$ powder was compared with the Bi$_2$O$_3$ powder and recognized as superior. The prepared BiF$_3$ powder sample was added in a plaster of Paris (POP) matrix in the proportion of 0%, 1%, 5%, and 10% by wt% to form POP–BiF$_3$(0%), POP–BiF$_3$(1%), POP–BiF$_3$(5%), and POP–BiF$_3$(10%) composite pellets, respectively, and activated the photocatalytic property under the UV–light irradiation,in the POP. In this work, Resazurin (Rz) ink was utilized as an indicator to examine the photocatalytic activity and self-cleaning performance of POP–BiF$_3$(0%), POP–BiF$_3$(1%), POP–BiF$_3$(5%), and POP–BiF$_3$(10%) composite pellets. In addition to the digital photographic method, the UV–visible absorption technique was adopted to quantify the rate of the de-colorization of the Rz ink, which is a direct measure of comparative photocatalytic performance of samples.

Keywords: BiF$_3$ nanostructure; POP composite; photocatalyst; Rz ink

1. Introduction

Recent development has focused on creating newly sustainable, low-cost photocatalytic materials with a superior performance than the traditional semiconductor photocatalysts such as TiO$_2$ and ZnO for self-cleaning applications [1]. In this direction, a huge potential is observed for the Bi-based semiconductors and their complexes [2–11]. This group of materials possess a direct band gap of a wide range from 2.5 eV to 3.2 eV and is severally reported for the fast rate of the creation of photo-induced charge carriers [12,13]. Moreover, most of all the Bi-based compounds show layered structures with polar surfaces and are found responsible for accelerating the separation of photo-generated carriers; consequently, reducing the recombination efficiency and showing a better photocatalytic activity even under the exposure of low intensity of light irradiation [13–17]. Among them, Bi$_2$O$_3$ has been studied extensively and found to be superior photocatalytic, highly photoconductive, and nontoxic in nature, having a narrow band gap of about 2.8 eV [15,16,18]. Bi$_2$O$_3$ has four kinds of polymorphs which are designated as α for the monoclinic structure, β for the tetragonal structure, γ for the body-centered cubic structure, and δ is for the face-centered cubic structure [19,20]. In several cases, Bi$_2$O$_3$ has been reported for dye degradation, the photosynthesis of organic compounds, and water splitting for hydrogen generation [21–23]. To enhance the photocatalytic performance further, the structure and the surfaces of the Bi$_2$O$_3$ compound are tailored extensively. For example, the β phase of

Bi_2O_3 is doped with Ti, which improves the photocatalytic activity comparatively [24]. In some cases, Bi_2O_3 has been further modified to form Bi_2O_3-based complexes with several other materials to create heterojunctions, which further supported the creation and separation of the photo-induced charge carriers in the heterointerface [25–28]. In previous reports, Singh et al. demonstrated the modification of Bi_2O_3 or Bi-based compounds through halogenations which lead to the formation of BiOCl and BiOF compounds, respectively, with a huge advancement in photocatalytic and self-cleaning properties [29,30].

Similarly, in the present work, Bi_2O_3 powder is used as the initial material and processed further with HF treatment and completely modified into the β phase of BiF_3. Usually, BiF_3 exists in two structures, cubic and orthorhombic, depicted as the α phase and β phase, respectively [31]. The α phase of BiF_3 and its applications are reported most commonly for simple synthesis methods and low cost with high photo-activity [32–34]. In a reported research work by Chenkai Feng et al., the α phase of the BiF_3 sample is prepared and then the photocatalytic performance is compared with the commercially available TiO_2 powder sample [32]. Interestingly, the α phase of BiF_3 is found 2.1 times superior to the TiO_2 powder sample [32]. However, the preparation and applications of BiF_3 having β phase are less explored. Therefore, in this work, after the preparation of the β phase of BiF_3, we analyzed its photocatalytic property and compared it with the initially purchased Bi_2O_3powder. In order to explore the possible utilization of BiF_3 for commercial application, it is important to look into a sustainable strategy. One of the methods could be in composite paints and coatings. BiF_3 can be physically mixed with any well-known and widely used materials, such as cement-based paints and other ceramic coatings. Plaster of Paris is known for the aesthetics and decoration material. It is also used in medicine to make casts for broken bones. To explore photocatalysis-based effects in plaster of Paris, it may be used for air cleaning as well as antibacterial properties. Hence, BiF_3 embedded in a plaster of Paris (POP) matrix is fabricated by varying the BiF_3 amounts from 0% to 10 wt%. Further, these POP-BiF_3(%) compositions are tested for the photocatalytic response on Resazurin (Rz) ink. Rz ink is used as a prototype carcinogenic pollutant and an indicator of photocatalytic performance.

2. Materials and Methods

In the process of making POP-BiF_3(%) composites, first, we prepared the powder of the BiF_3 sample. We obtained a precipitation technique to prepare the BiF_3 sample. The fixed amount (5 g) of pure Bi_2O_3 powder of AR grade (Hi-media, ≥99%) was dissolved in the hydrofluoric (HF) acid (Qualikems 40%, Vadodara, India) solution (30 mL) and stirred for 30 min. Under the stirring with HF solution, the yellow color of Bi_2O_3 gradually changed into white-color powder. The product of white powder was washed in distilled water several times followed by acetone and dried at 80 °C for 24 h in a vacuum oven. A fraction of the sample was collected for testing, named sample Bi_2O_3-HF-1. Again, the remaining product of white powder of Bi_2O_3-HF-1 sample was dissolved in the concentrated HF solution and stirred for 30 min. The output product obtained after the second treatment from HF solution was followed with the same procedures of washing, filtering, and drying as for the sample Bi_2O_3-HF-1, and we procured the test sample2 named as sample Bi_2O_3-HF-2. Similarly, a test sample3 was procured and named as sample Bi_2O_3-HF-3 for further testing.

Next, by using the Bi_2O_3-HF-3 powder sample, we prepared POP-BiF_3(%) composite pellets. In the composites, we maintained the concentration of Bi_2O_3-HF-3 powder in the POP matrix in accordance with 0, 1, 5, and 10 by wt%. With respect to each composition of POP-BiF_3(%) composites, the calculated amount of Bi_2O_3-HF-3 powder sample and POP were mixed rigorously to prepare a homogeneous mixture, separately. The paste of each mixture of different POP-BiF_3(%) composites was obtained by adding an equal amount of distilled water. Individually, the paste of different POP-BiF_3(%) composites was transferred into cylindrical molds of 20 mm diameter and 10 mm height to prepare the pellets of each composition, respectively. Finally, the pellets were left to naturally dry fortwo days.

The structural analysis of Bi_2O_3, HF-treated Bi_2O_3 test samples, and POP-BiF_3(%) composite pellets was performed through X-ray diffraction (XRD) (Rigaku), having 9 kW rotating anode and Cu Kα source. Microstructure analysis was obtained from FE-SEM (Inspect™S50). Optical property and photocatalytic performance were tested via UV–visible spectrophotometer of double beam (Thermo Scientific, Evolution 220, Waltham, MA, USA). In addition, to carry out the photocatalytic reaction, we used a box inbuilt with a lamp (Hitachi FL8BL-Blight) as a UV light source having maximum emission ~355 nm wavelengths. The distance between the lamps and the samples was adjusted such that the intensity falling on the samples was maintained at about ~3200 lx.

3. Results and Discussion

The systematic XRD results of Bi_2O_3 and the samples obtained from the successive fluorination of Bi_2O_3 via HF solution are shown in Figure 1. The step-wise fluorinated samples are denoted as Bi_2O_3-HF-1, Bi_2O_3-HF-2, and Bi_2O_3-HF-3, respectively. XRD of the purchased Bi_2O_3 sample was compared with the JCPDS file no 76–1730 and matched with the monoclinic phase. XRD results of the Bi_2O_3-HF-1 sample revealed that after the 1st washing of Bi_2O_3 powder, new diffraction peaks appeared in the X-ray diffraction pattern. A set of these new diffraction peaks was related to the intermediate phase of $Bi_{1.2}F_{2.4}O_{0.6}$(PDF-36-0457), which are marked as '*'. The other set of remaining peaks with less intensity belongs to the orthorhombic structure of BiF_3 (PDF-70-2407). XRD results of the Bi_2O_3-HF-2 sample for a 2nd consecutive washing of the Bi_2O_3 powder showed the relative increase in the intensity of diffraction peaks belonging to the BiF_3 (PDF-70-2407) phase structure at the expense of the diffraction peaks belonging to an intermediate phase of $Bi_{1.2}F_{2.4}O_{0.6}$ (PDF-36-0457), relatively. The almost pure phase of BiF_3 (PDF-70-2407) appeared after the 3rd consecutive washing of Bi_2O_3 powder in addition to a very small quantity of an unidentified impure phase which is marked as '#'. Thus, multiple washing of Bi_2O_3 powder through the concentrated HF solution led towards the formation of the almost pure orthorhombic structure of the BiF_3 powder sample.

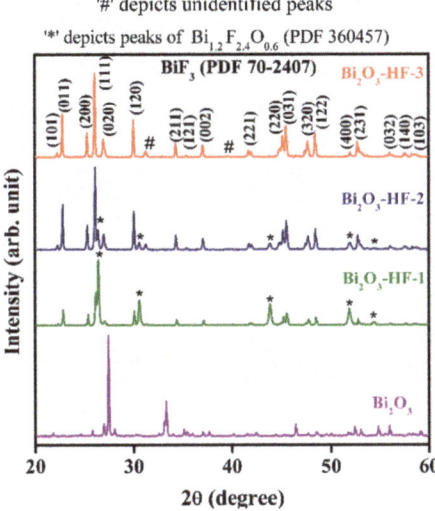

Figure 1. X-ray diffraction patterns of purchased Bi_2O_3 and Bi_2O_3 after three successive HF treatments depicted as Bi_2O_3-HF-1, Bi_2O_3-HF-2, Bi_2O_3-HF-3, and Bi_2O_3-HF-4, respectively.

Only a few solution techniques have been used for the formations of various phases of the BiF_3 sample through different methods. For example, Feng et al. reported the formation of BiF_3 (JCPDS: 51-0944) by a simple water-bath method, where they kept the molar ratio of

Bi and F above 1:3, otherwise the impurity of Bi_2O_3 and BiOF remained present [32]. In this method, the constituent of Bi was obtained from Bi_2O_3 while the element of F was attained from the NH_4F solution [32]. Zhao et al. reported the evolution of the BiF_3 nanocrystals in various shapes from monodispersed nano-plates to nano-rods and then to nanowires via the novel acid–base couple extraction route and tuning the molar ratio of F vs. Bi [31]. Sarkar et al. used Poly (vinyl pyrrolidone) (PVP) for the encapsulation and formation of cubic nanocrystals of BiF_3 via the hydrothermal method [35]. In another method, by using a novel ion-exchange approach, Kan et al. produced pure BiF_3 (JCPDS: 73-1988). Here, the NH_4F solution was used for the constituent of F while BiOCl was utilized to attain the element of Bi in accordance with the molar ratio (RF = F/Bi) of 8:1 [36]. Below to the molar ratio of 8:1 (RF), the final product consisted of a small amount of $Bi_7F_{11}O_5$ as an impurity phase [36].

Contrary to the above reported studies, in the present method, the ratio of O/F was controlled via a chemical bath of Bi_2O_3 in a concentrated HF solution. The constituent of Bi was extracted from the powder of Bi_2O_3, while for the element of F in a concentrated HF solution was utilized. Bi_2O_3 powder was washed several times from the concentrated HF solution, which may have led to two types of products, as follows in the reaction mechanisms one and two given below:

$$2Bi^{3+} + 6F^- \rightarrow 2BiF_3 \rightarrow \text{(Concentrated HF)} \quad (1)$$

$$2Bi^{3+} + 2H_2O + 6F^- \rightarrow 2BiOF\downarrow + 4HF \text{ (Diluted HF)} \quad (2)$$

Multiple washing from HF solution increased the constituent of F and led towards the formation of BiF_3 from Bi_2O_3 powder. Normally, if the Bi_2O_3 powder is washed from the concentrated HF solution, the positively charge Bi^{3+} ion reacts with F^- ion and forms BiF_3 and follows reaction mechanism one. However, in the HF solution, some water content always remains present; therefore, in the case of the 1st washing, some of the Bi_2O_3 powder converted into BiF_3 according to reaction mechanism one, while some of the Bi_2O_3 powder reacted with HF as well as the water content, followed reaction mechanism2 and formed an intermediate product of the BiOF family, which was recognized as $BiO_{0.51}F_{1.98}$ in the present case. Further, the 2nd and 3rd washing provided more and more F^- ion in the solution which again reacted with the intermediate product of $BiO_{0.51}F_{1.98}$ and converted it to the final product of the BiF_3 sample. Multiple washing and high concentrations of HF increased the F concentration in the O/F ratio and led toward the cubic-αBiF_3 phase (PDF-073-1988) from the Bi_2O_3 powder.

Further, the BiF_3 powder was investigated for its photocatalytic performance and compared with the Bi_2O_3 powder sample. A total of 0.05 g powder of both Bi_2O_3 and BiF_3 were sonicated in 50-50 mL water solutions of the hazardous dye of Methylene blue (MB), separately, as test solutions. Before the photocatalytic investigation of Bi_2O_3 and BiF_3 samples, to neglect the effect of the adsorption–desorption of the dye over the surfaces of these powders, first both the test solutions of Bi_2O_3 and BiF_3 in MB were sonicated under dark for a 30 min duration to achieve the adsorption–desorption equilibrium. Then, both the test solutions were transferred under the UV light irradiation of a 355 nm wavelength. To probe the photocatalytic activity of the Bi_2O_3 and BiF_3 powder samples, 1-1 mL of the MB was collected in a separate Eppendorf from both the test solutions at fixed time intervals. Finally, the absorption study of the collected samples was carried out by using a UV–visible spectrophotometer. After the fixed time of reaction, both the solutions were sonicated for five minutes to maintain the homogeneity and, then, 1-1 mL of the MB was collected. Each collected sample was centrifuged to remove the segregated residual from the solutions. At last, the photocatalytic decomposition of MB for each collected sample from both the set of solutions immersed with Bi_2O_3 and BiF_3 powders were tested from UV–visible spectroscopy, respectively.

The adsorption–desorption reaction analysis for both the samples under the dark environment showed an insignificant change in the concentration of MB solutions (absorbance results probed by UV–visible spectroscopy is not shown here). After adsorption–desorption, the photolysis and photocatalytic results of decomposition, as well as the kinetic rate of the decomposition of MB solutions under UV exposure, were plotted and represented in Figure 2a–c. The decomposition of the MB solution probed via absorbance data of UV-visible spectroscopy confirmed a fast degradation of MB solution for the BiF_3 powder sample as compared to the Bi_2O_3 powder. For comparison purposes, and to remove the artifacts during the photocatalytic reaction, the photolysis of the blank MB solution was obtained for the same parameters and for the same time scale as maintained for the photocatalytic decomposition of MB solutions for the Bi_2O_3 and BiF_3 powder samples. A control reaction of photolysis of blank MB solution under the exposure of 355 nm light radiation showed a negligible change in the concentration. In Figure 2d,e, the Co stands for initial concentration, while Ct represents the time-dependent concentration of the MB solution after the photocatalytic decomposition. Using the parameters Co and Ct and their relation with time, i.e., $\ln(C/Co) = k_f\, t$, the decomposition rate constant of MB solutions due to the photocatalytic reaction for both the powder samples were calculated and compared. Here, k_f represents the rate constant of photocatalytic reactions, which was calculated as 0.0011 and 0.0103 min^{-1} for the Bi_2O_3 and BiF_3 powders, respectively. Thus, the photocatalytic degradation of the BiF_3 powder sample was confirmed to be multiple times higher than the Bi_2O_3 powder sample. A comprehensive study of photocatalysis is shown in Table 1.

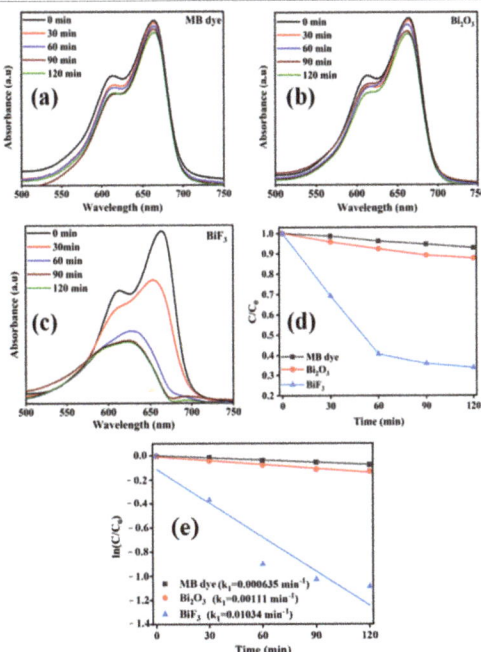

Figure 2. Absorption vs. wavelength spectra for (a) the photolysis of Methylene blue (MB), and the photocatalysis of MB under the exposure of UV light of 355 nm wavelength due to (b) Bi_2O_3 powder and (c) BiF_3 powder (d) respective C/Co vs. time plots, and (e) $\ln(C/Co)$ vs. time plots to obtain the pseudo-first-order reaction rate constant for the photolysis of MB and the photocatalysis of MB.

Four different composite pellets of POP–BiF_3(%) were obtained. The BiF_3 power was varied in the POP matrix in accordance with 0, 1, 5, and 10 by wt%. Here, the pellets of POP–BiF_3(%) composites were named as POP–BiF_3(0%), POP–BiF_3(1%), POP–BiF_3(5%), and POP–BiF_3(10%), respectively. The composite formation and purity of POP–BiF_3(0%),

POP–BiF$_3$(1%), POP–BiF$_3$(5%), and POP–BiF$_3$(10%) pellets were checked and verified through XRD (not shown here).

Again, the absorption of UV–visible radiation for POP–BiF$_3$(0%), POP–BiF$_3$(1%), POP–BiF$_3$(5%), and POP–BiF$_3$(10%) pellets was probed and utilized for the bandgap calculations according to the Kubelka–Munk model [37]. Results are shown in Figure 3a,b. Figure 3a demonstrates the gradual increase in the absorption from the visible to UV region for the POP sample. Adding BiF$_3$ in the POP matrix improved the absorption towards the visible range, relatively. The results of the absorption coefficient (α) vs. incident photon energy E (hυ) obtained from Figure 3a were extrapolated according to the Kubelka–Munk model to calculate the direct bandgap of POP and POP–BiF$_3$(%) composites, as shown in Figure 3b. Results showed that the bandgap tended to decrease with the addition of BiF$_3$ in the POP matrix from 3.69 eV to 3.26 eV, respectively.

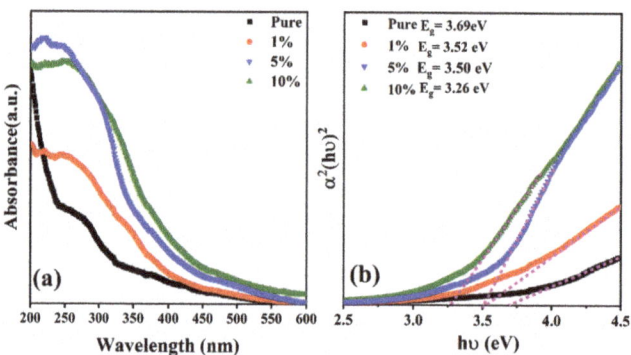

Figure 3. (a) Absorption spectra and (b) $\alpha^2(h\upsilon)^2$ vs. hυ plots for the band gap (E_g) calculation of POP–BiF$_3$ (%) composites with the variation of BiF$_3$ concentration 0%, 1%, 5%, and 10% in POP matrix.

The variation in morphology of the POP–BiF$_3$(%) composite pellets along with the change of BiF$_3$ concentration in the POP matrix were probed through SEM and are shown in Figure 4a–d. The surface image of POP–BiF$_3$(0%) pellet stands for the pure POP sample and shows a homogeneous rod-shape microstructure. These rods were entangled with each other and made net-like porous surfaces. One end of the rod was defused in the surface while the other end seemed to emerge from the surface and dangle. Such entangled rods over the surface of the POP–BiF$_3$(0%) pellet were due to water treatment and the continuous hydration of POP which led to the continuous nucleation and growth of rods in random orientations. SEM images of the BiF$_3$-added POP–BiF$_3$(0%), POP–BiF$_3$(1%), POP–BiF$_3$(5%), and POP–BiF$_3$(10%) composite pellets revealed almost similar surfaces to the POP–BiF$_3$(0%) pellet accompanied with several entangled random oriented rods forming the net-like porous structure. However, with the addition of BiF$_3$ in the POP matrix, the porosity of the surface reduced in POP–BiF$_3$(1%), POP–BiF$_3$(5%), and POP–BiF$_3$(10%) composite pellets, relatively, as compared to the surface of the POP–BiF$_3$(0%) pellet. The addition of BiF$_3$ may have filled the pores of POP and, therefore, led to a smooth and glassy surface in the POP–BiF$_3$(10%) composite pellet.

Further, to check the photocatalytic performance of BiF$_3$ in the POP matrix, the Rz ink was prepared by using a known reported method [38]. Rz ink is an indicator of self-cleaning/photocatalysis, which changes its color if coated over any surface of photocatalytic material under the exposure of UV irradiation. Therefore, an equal amount of ink was pasted over each surface of POP–BiF$_3$(0%), POP–BiF$_3$(1%), POP–BiF$_3$(5%), and POP–BiF$_3$(10%) composite pellets. These pellets were subjected to UV light (355 nm) illumination with equal exposures and monitored with time on several fixed intervals for 180 min of time duration. Each time, surface images of POP–BiF$_3$(0%), POP–BiF$_3$(1%),

POP–BiF$_3$(5%), and POP–BiF$_3$(10%) composite pellets were obtained via a high-quality digital camera. For comparison purposes, all captured images on certain time intervals were compiled together and shown in Figure 5. In correlation to Figure 5, ΔRGB't vs. time is plotted and shown as Figure 6a,b, which represents the change in blue as Figure 6a and change in red color as Figure 6b due to photo-reduction in the Rz indicator ink under the UV light irradiation on POP–BiF$_3$(%) composites, respectively.

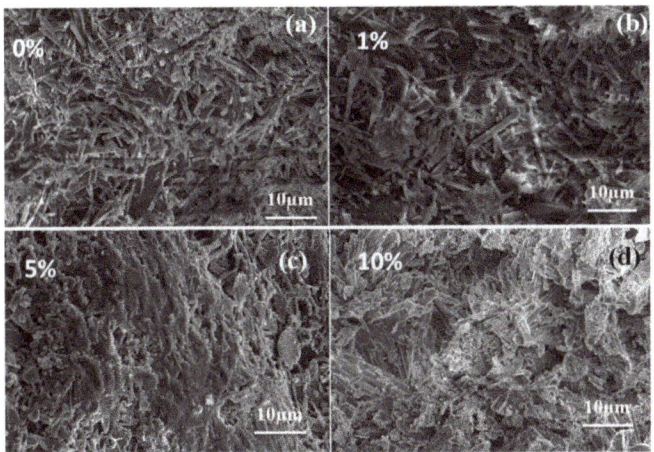

Figure 4. Typical SEM images of POP–BiF$_3$ (%) composites along with the concentration variation of BiF$_3$ for (**a**) 0%, (**b**) 1%, (**c**) 5%, and (**d**) 10% in POP matrix.

Table 1. Comprehensive study of photocatalysis activity of BiF$_3$ catalyst.

Catalyst	Process	Pollutant	Power Source	Catalysis Time	Performance
BiOI/BiF$_3$ composite [39]	Photocatalysis	Tetracycline hydrochloride	Visible light	120 min	~75.6%
BiF$_3$–Bi$_2$NbO$_5$F core–shell [34]	Photocatalysis	RB dye	Visible light	90 min	0.028 min^{-1}
BiOCl/BiF$_3$ heterojunction [40]	Photocatalysis	MO dye	UV light	30 min	~90%
BiF$_3$ octahedrons [41]	Photocatalysis	RB dye	Solar light	50 min	~95.7%
BiF$_3$ nanoparticles [32]	Photocatalysis	RB dye	UV light	50 min	~78.5%
BiF$_3$/BiOBr heterojunctions [42]	Photocatalysis	MO dye	Visible light	200 min	~82.6%
Bismuth Fluoride Surface Crystallized 2Bi$_2$O$_3$-B$_2$O$_3$ Glass [43]	Photocatalysis	Rhodamine 6G	Halogen lamp	120 min	~85%
Bismuth Fluoride on SrO-Bi$_2$O$_3$-B$_2$O$_3$ transparent Glass ceramic [44]	Photocatalysis	MB dye	Visible lights	540 min	0.02226 min^{-1}
BiOCl/BiF$_3$ on ZnO-Bi$_2$O$_3$-B$_2$O$_3$ glass [45]	Photocatalysis	MB dye	UV light	300 min	~90%
BiF$_3$ (present study)	Photocatalysis	MB dye	UV light	120 min	0.0103 min^{-1}

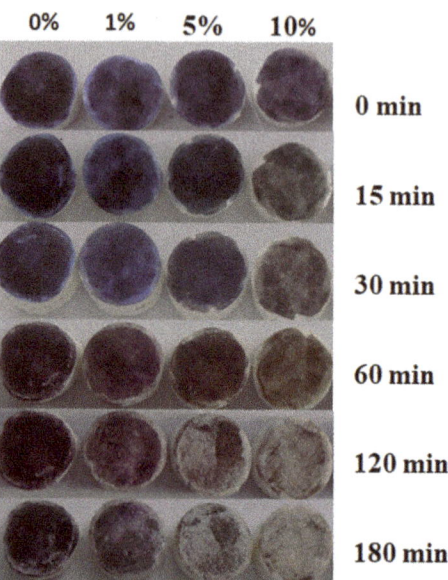

Figure 5. Photo-reduction in Rz indicator ink for 3 h reaction timeline on POP–BiF$_3$ (%) composites having the variation of BiF$_3$ 0%, 1%, 5%, and 10% by wt., respectively, under the exposure of 355 nm wavelength of UV light irradiation.

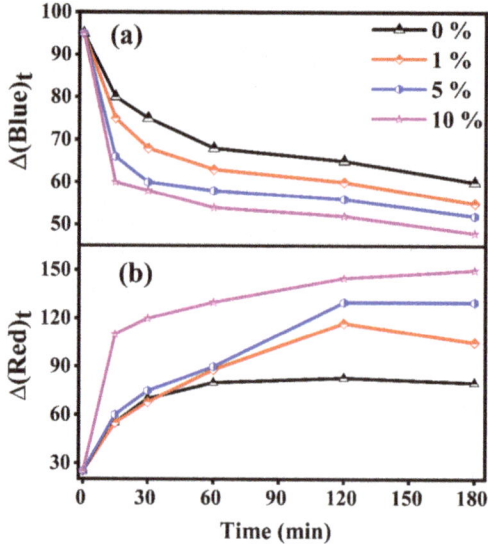

Figure 6. ΔRGB't vs. time plots for (**a**) blue and (**b**) red color change due to photo-reduction in Rz indicator ink on POP–BiF$_3$(%) composites having BiF$_3$ 0%, 1%, 5%, and 10% by wt., respectively, under the UV light irradiation.

In the present case, under the exposure of UV light on the POP–BiF$_3$(0%) composite pellet, which had no BiF$_3$ content, showed no color change on the surface up to the 30 min of time duration, while a slight change after 60 min to 180 min was monitored. This slight change from a blue to purple color was due to the photolysis of the Rz ink. For POP–

BiF$_3$(1%), POP–BiF$_3$(5%), and POP–BiF$_3$(10%) composite pellets, the color of the Rz ink readily changed in proportion to the BiF$_3$ content in POP from royal blue to pink and then into colorless ink. The rate of change in the color of the Rz ink was systematic and increased for POP–BiF$_3$(0%), POP–BiF$_3$(1%), POP–BiF$_3$(5%), and POP–BiF$_3$(10%) composite pellets, respectively. The color of the Rz ink over the surface of the POP–BiF$_3$(10%) composite pellet appeared almost colorless within the 180 min time period.

Usually, the rate of the change of color of the Rz ink indicates the rate of photocatalytic reduction. The reaction mechanism of photocatalytic reduction and the color-change of the Rz ink are mentioned as under Equations (3)–(6).

$$BiF_3 \xrightarrow{h\vartheta} BiF_3(e^-, h^+) \quad (3)$$

$$Glycerol \xrightarrow{h^+} OH + Glyceric\ acid \quad (4)$$

$$\text{Resorufin (Rz) (Blue)} \xrightarrow{\cdot OH} \text{Resorufin (Rf) (Pink)} \quad (5)$$

$$BiF_3 * (e^-, h^+) \rightarrow BiF_3 \quad (6)$$

At the same time of the image capture, the POP–BiF$_3$(0%), POP–BiF$_3$(1%), POP–BiF$_3$(5%), and POP–BiF$_3$(10%) composite pellets were again tested through UV–visible spectroscopy for a quantitative analysis of the photocatalytic reduction in the Rz ink coated over the surfaces. Absorption peak intensities corresponding to the photo catalytic reduction were monitored at two different wavelengths, i.e., 630 nm and 581 nm. The absorption spectra obtained from the UV–visible spectroscopy is illustrated in Figure 7a–d. The peak intensity monitored at the 630 nm wavelength directly correlated to the blue color of the Rz ink and with the exposure of UV light (355 nm). The peak intensity of absorbance decreased at the wavelength of 630 nm along with the color change from blue to pink, resulting in the formation of resorufin (Rf) as a byproduct. The intensity decay of another peak at 581 nm represented the photobleaching of the Rf molecule as and when the color changed from pink to colorless. Under the effect of UV light irradiation, a negligible change in the intensity of characteristic absorption of the Rz ink was observed for the POP–BiF$_3$(0%) composite pellet; however, the absorption intensity of the Rz ink decreased consistently for all the other samples just in accordance with the color change observed in the digital photographs as shown in Figure 5. The absorption results are shown in Figure 7a–d was further utilized to extract the kinetic rate of photocatalytic reduction and photo mineralization of intermediates of the Rz ink monitored for both the wavelengths at 630 nm and 581 nm, respectively, shown in Figure 8. In Figure 8, Co represents the initial (t = 0) absorbance of the Rz ink and C is the absorbance of the Rz ink, which varied with time t. The Co and C and kinetic rate of photocatalytic reduction was calculated corresponding to the absorption spectra monitored for both the wavelengths of absorbance, i.e., 581 nm and 630 nm of Rz ink. For the POP–BiF$_3$(0%) sample, the results showed that the intensity decay, as well as the kinetic rate of reaction due to the photocatalytic degradation of the Rz ink corresponding to the wavelengths monitored at 581 nm and 630 nm for 180 min of time duration, was less. Generally, the photocatalytic reaction rate was monitored to correspond faster to the absorbance at 630 nm wavelengths in comparison to the absorbance around the 581 nm wavelength for all samples. As compared to POP–BiF$_3$(0%), the POP–BiF$_3$(1%), POP–BiF$_3$(5%), and POP–BiF$_3$(10%) composite pellets showed relatively faster kinetic rate of photocatalytic reduction in the Rz ink, respectively.

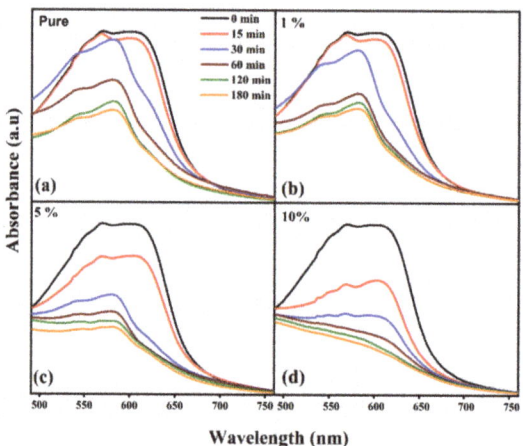

Figure 7. Absorption vs. wavelength spectra for the photocatalysis of Rz ink on the surface of POP–BiF$_3$(%) composites with the variation of BiF$_3$ concentration (**a**) 0%, (**b**) 1%, (**c**) 5%, and (**d**) 10% in POP matrix.

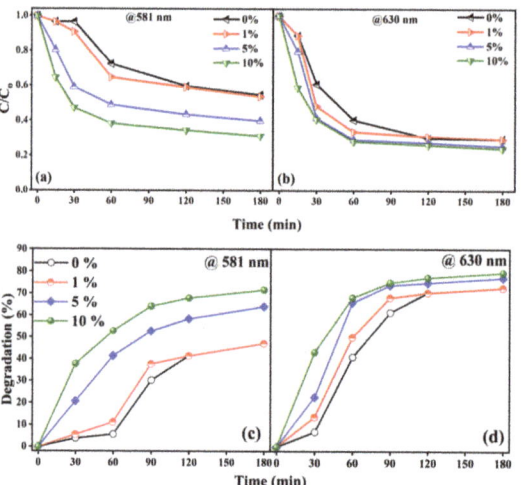

Figure 8. The C/Co vs. time plots for photocatalytic degradation of Rz ink for POP–BiF$_3$(%) composites monitored at: (**a**) 581 nm and (**b**) 630 nm wavelength, and respective percentage of photocatalytic degradation for (**c**) 581 nm and (**d**) 630 nm wavelength.

For simplicity, the photoreduction due to UV light irradiation of Rz ink, which was coated on POP–BiF$_3$ composite pellets, is shown in Figure 9. Under the illumination of UV light irradiation, the catalyst material BiF$_3$ present in POP generated sacrificial donor electrons on its surface which reacted with the photo-induced holes of the Rz ink. Therefore, the glycerol present in the Rz ink utilized this sacrificial donor electron and generated an •OH radicle along with glyceric acid as a by-product. At the same time, these intermediate •OH radicles reduced the blue color of the Rz ink into Rf of the pink color after the long-time illumination of UV light irradiation, resulting in the reduction in the Rf molecule into a colorless product. Similar to the Rz ink, the harmful pollutants may have also reduced into non-harmful products with the aid of hydroxyl radicals generated on the surface of the photocatalytic BiF$_3$ in the POP matrix. In the present work, we obtained a maximum content of BiF$_3$ only up to 10% in the POP matrix. The Further addition of BiF$_3$ in POP may have led to the decay in the mechanical strength of the architecture. The concentration of BiF$_3$ inside the POP matrix along with an optimized photocatalytic performance and superior strength of the structures are a further matter of research. Thus, the above results demonstrate the successful photocatalytic application of BiF$_3$ inside the POP matrix. POP is an important cementitious material used for the making of several building constructions and sculptures. We propose here for the addition of BiF$_3$ as an activated photocatalytic material in the side of any cementitious material for self-cleaning of building constructions and a reduction in environmental pollution.

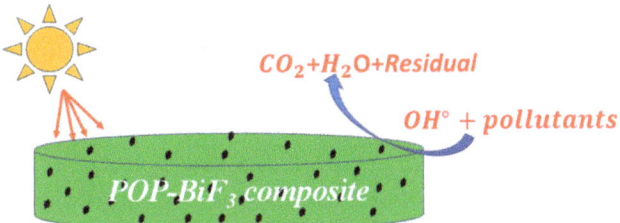

Figure 9. The schematic representation of the proposed mechanism for the photocatalytic degradation of pollutants under UVlight irradiation over POP–BiF$_3$ composite pellet surfaces.

4. Conclusions

Successfully, theBiF$_3$ powder sample was prepared via the precipitation route simply by washing the purchased powder of Bi$_2$O$_3$ several times into the concentrated HF solution. After several washes through the concentrated HF solution, the Bi$_2$O$_3$ powder was systematically transformed into BiF$_3$ powder in conjunction with each washing. An intermediate phase of Bi$_{1.2}$F$_{2.4}$O$_{0.6}$ was identified for the 1st and 2nd washing of the Bi$_2$O$_3$ powder, which completely vanished after the 3rd washing and converted into the BiF$_3$ powder sample. The photocatalytic performance of the as-prepared BiF$_3$ powder was tested and compared with the Bi$_2$O$_3$ powder on a hazardous industrial waste solution of MB dye. The BiF$_3$ powder rapidly decomposed the MB solution under the visible light illumination as compared to the purchased Bi$_2$O$_3$ powder sample. Usually, POP is used vastly in the construction as well as the building of sculptures and, therefore, is always exposed to water as well as air pollutants. Therefore, to check the self-cleaning activity from such pollutants and introduce the photocatalytic properties in POP, BiF$_3$ powders were added according to 0%, 1%, 5%, and 10% by wt% in the POP matrix. The photocatalytic and self-cleaning properties of these POP–BiF$_3$ composites were demonstrated by using a well-known photocatalysis indicator ink of Rz under visible light exposure. Due to the photocatalytic effect of POP–BiF$_3$ composites under the solar illumination, the blue color of the Rz ink turned into Rf of the pink color. The long-time illumination of visible light resulted from the reduction in the pink color Rf molecules into the colorless product. Photocatalytic performances were improved linearly for the incorporation of BiF$_3$ up to 10% of the concentration in the POP matrix.

Author Contributions: Conceptualization, V.P.S. and R.V.; methodology, V.P.S., R.V., S.-H.P. and M.K.; validation, V.P.S., M.S. and D.M.; formal analysis, V.P.S., M.S. and D.M.; investigation, V.P.S., M.K., M.S. and D.M.; resources, R.V., K.-S.S. and S.-H.P.; data curation, V.P.S. and M.K.; writing, V.P.S., R.V., S.-H.P. and M.K.; writing—review and editing, V.P.S., R.V., S.-H.P. and M.K.; visualization, V.P.S., R.V., S.-H.P. and M.K.; supervision, R.V., K.-S.S. and S.-H.P.; project administration, R.V., K.-S.S. and S.-H.P.; funding acquisition, K.-S.S. and S.-H.P. All authors have read and agreed to the published version of the manuscript.

Funding: This work was supported by the 2021 Yeungnam University Research Grant.

Acknowledgments: V. P. Singh and Rahul Vaish acknowledge the NPIU and AICTE for the CRS grant under the TEQIP III program for research-related work. RV thanks the CSIR, New Delhi, India, for the financial support under the sponsored research project scheme.

Conflicts of Interest: The authors declare no conflict of interest.

References

1. Li, J.; Yu, Y.; Zhang, L. Bismuth oxyhalide nanomaterials: Layered structures meet photocatalysis. *Nanoscale* **2014**, *6*, 8473–8488. [CrossRef]
2. Tian, G.; Chen, Y.; Zhou, W.; Pan, K.; Dong, Y.; Tian, C.; Fu, H. Facile solvothermal synthesis of hierarchical flower-like Bi_2MoO_6 hollow spheres as high performance visible-light driven photocatalysts. *J. Mater. Chem.* **2011**, *21*, 887–892. [CrossRef]
3. Zhang, B.L.W.; Wang, Y.J.; Cheng, H.Y.; Yao, W.Q.; Zhu, Y.F. Synthesis of porous Bi_2WO_6 thin films as efficient visible-light-active photocatalysts. *Adv. Mater.* **2009**, *21*, 1286–1290. [CrossRef]
4. Li, R.; Zhang, F.; Wang, D.; Yang, J.; Li, M.; Zhu, J.; Zhou, X.; Han, H.; Li, C. Spatial separation of photogenerated electrons and holes among {010} and {110} crystal facets of $BiVO_4$. *Nat. Commun.* **2013**, *4*, 1432. [CrossRef]
5. Wang, B.C.; Nisar, J.; Pathak, B.; Kang, T.W.; Ahuja, R. Band gap engineering in $BiNbO_4$ for visible-light photocatalysis. *Appl. Phys. Lett.* **2012**, *100*, 182102. [CrossRef]
6. Shi, R.; Lin, J.; Wang, Y.; Xu, J.; Zhu, Y. Visible-light photocatalytic degradation of $BiTaO_4$ photocatalyst and mechanism of photocorrosion suppression. *J. Phys. Chem. C* **2010**, *114*, 6472–6477. [CrossRef]
7. You, Q.; Fu, Y.; Ding, Z.; Wu, L.; Wang, X.; Li, Z. A facile hydrothermal method to $BiSbO_4$ nanoplates with superior photocatalytic performance for benzene and 4-chlorophenol degradations. *Dalton Trans.* **2011**, *40*, 5774–5780. [CrossRef] [PubMed]
8. Wu, J.; Huang, F.; Lü, X.; Chen, P.; Wan, D.; Xu, F. Improved visible-light photocatalysis of nano-$Bi_2Sn_2O_7$ with dispersed s-bands. *J. Mater. Chem.* **2011**, *21*, 3872–3876. [CrossRef]
9. Gao, F.; Chen, X.; Yin, K.; Dong, S.; Ren, Z.; Yuan, F.; Yu, T.; Zou, Z.G.; Liu, J.-M. Visible-light photocatalytic properties of weak magnetic $BiFeO_3$ nanoparticles. *Adv. Mater.* **2007**, *19*, 2889–2892. [CrossRef]
10. Zhang, Q.; Gong, W.; Wang, J.; Ning, X.; Wang, Z.; Zhao, X.; Ren, W.; Zhang, Z. Size-dependent magnetic, photoabsorbing, and photocatalytic properties of single-crystalline $Bi_2Fe_4O_9$ semiconductor nanocrystals. *J. Phys. Chem. C* **2011**, *115*, 25241–25246. [CrossRef]
11. Pan, C.; Zhu, Y. Size-controlled synthesis of $BiPO_4$ nanocrystals for enhanced photocatalytic performance. *J. Mater. Chem.* **2011**, *21*, 4235–4241. [CrossRef]
12. Lin, X.; Huang, F.; Wang, W.; Shi, J. Photocatalytic activity of $Bi_{24}Ga_2O_{39}$ for degrading methylene blue. *Scr. Mater.* **2007**, *56*, 189–192. [CrossRef]
13. Mohn, C.E.; Stølen, S. Influence of the stereochemically active bismuth lone pair structure on ferroelectricity and photocalytic activity of Aurivillius phase Bi_2WO_6. *Phys. Rev. B Condens. Matter Mater. Phys.* **2011**, *83*, 01410. [CrossRef]
14. Yu, C.; Yang, P.; Tie, L.; Yang, S.; Dong, S.; Sun, J.; Sun, J. One-pot fabrication of β-Bi_2O_3@Bi_2S_3 hierarchical hollow spheres with advanced sunlight photocatalytic RhB oxidation and Cr(VI) reduction activities. *Appl. Surf. Sci.* **2018**, *455*, 8–17. [CrossRef]
15. Yan, Q.; Xie, X.; Liu, Y.; Wang, S.; Zhang, M.; Chen, Y.; Si, Y. Constructing a new Z-scheme multi-heterojunction photocataslyts Ag-AgI/BiOI-Bi_2O_3 with enhanced photocatalytic activity. *J. Hazard. Mater.* **2019**, *371*, 304–315. [CrossRef] [PubMed]
16. Chen, L.; He, J.; Yuan, Q.; Liu, Y.; Au, C.T.; Yin, S.F. Environmentally benign synthesis of branched Bi_2O_3-Bi_2S_3 photocatalysts by an etching and re-growth method. *J. Mater. Chem. A* **2015**, *3*, 1096–1102. [CrossRef]
17. Zou, S.; Teng, F.; Chang, C.; Liu, Z.; Wang, S. Controllable synthesis of uniform BiOF nanosheets and their improved photocatalytic activity by an exposed high-energy (002) facet and internal electric field. *RSC Adv.* **2015**, *5*, 88936–88942. [CrossRef]
18. Lu, H.; Hao, Q.; Chen, T.; Zhang, L.; Chen, D.; Ma, C.; Yao, W.; Zhu, Y. A high-performance Bi_2O_3/Bi_2SiO_5 p-n heterojunction photocatalyst induced by phase transition of Bi_2O_3. *Appl. Catal. B Environ.* **2018**, *237*, 59–67. [CrossRef]
19. Bian, Z.; Zhu, J.; Wang, S.; Cao, Y.; Qian, X.; Li, H. Self-Assembly of Active Bi_2O_3/TiO_2 Visible Photocatalyst with Ordered Mesoporous Structure and Highly Crystallized Anatase. *J. Phys. Chem. C* **2008**, *112*, 6258–6262. [CrossRef]
20. Hameed, A.; Montini, T.; Gombac, V.; Fornasiero, P. Surface phases and photocatalytic activity correlation of Bi_2O_3/Bi_2O_{4-x} nanocomposite. *J. Am. Chem. Soc.* **2008**, *130*, 9658–9659. [CrossRef]
21. Brezesinski, K.; Ostermann, R.; Hartmann, P.; Perlich, J.; Brezesinski, T. Exceptional photocatalytic activity of ordered mesoporous β-Bi_2O_3 thin films and electrospun nanofiber mats. *Chem. Mater.* **2010**, *22*, 3079–3085. [CrossRef]
22. Dai, Y.; Yin, L. Synthesis and photocatalytic activity of Ag-Ti-Si ternary modified α-Bi_2O_3 nanoporous spheres. *Mater. Lett.* **2015**, *142*, 225–228. [CrossRef]
23. Naik, B.; Parida, K.M.; Behera, G.C. Facile synthesis of Bi_2O_3/$TiO_{2-x}N_x$ and its direct solar-light-driven photocatalytic selective hydroxylation of phenol. *ChemCatChem* **2011**, *3*, 311–318. [CrossRef]
24. Yin, L.; Niu, J.; Shen, Z.; Chen, J. Mechanism of reductive decomposition of pentachlorophenol by ti-doped β-Bi_2O_3 under visible light irradiation. *Environ. Sci. Technol.* **2010**, *44*, 5581–5586. [CrossRef] [PubMed]
25. Zhao, W.; Zhang, J.; Zhu, F.; Mu, F.; Zhang, L.; Dai, B.; Xu, J.; Zhu, A.; Sun, C.; Leung, D.Y. Study the photocatalytic mechanism of the novel Ag/p-Ag_2O/n-$BiVO_4$ plasmonic photocatalyst for the simultaneous removal of BPA and chromium(VI). *Chem. Eng. J.* **2019**, *361*, 1352–1362. [CrossRef]

26. Habibi, Y.A.; Mousavi, M.; Nakata, K. Boosting visible-light photocatalytic performance of g-C$_3$N$_4$/Fe$_3$O$_4$ anchored with CoMoO$_4$ nanoparticles: Novel magnetically recoverable photocatalysts. *J. Photochem. Photobiol. A Chem.* **2019**, *368*, 120–136. [CrossRef]
27. Habibi, Y.A.; Mousavi, M. Deposition of CuWO$_4$ nanoparticles over g-C$_3$N$_4$/Fe$_3$O$_4$ nanocomposite: Novel magnetic photocatalysts with drastically enhanced performance under visible-light. *Adv. Powder Technol.* **2018**, *29*, 1379–1392. [CrossRef]
28. Mousavi, M.; Habibi, Y.A.; Pouran, S.R. Review on magnetically separable graphitic carbon nitride-based nanocomposites as promising visible-light-driven photocatalysts. *J. Mater. Sci. Mater. Electron.* **2018**, *29*, 1719–1747. [CrossRef]
29. Singh, V.P.; Mishra, D.; Kabachkov, E.N.; Shul'ga, Y.M.; Vaish, R. The characteristics of BiOCl/Plaster of Paris composites and their photocatalytic performance under visible light illumination for self-cleaning. *Mater. Sci. Energy Technol.* **2020**, *3*, 299–307. [CrossRef]
30. Singh, V.P.; Vaish, R. Hierarchical growth of BiOCl on SrO-Bi$_2$O$_3$-B$_2$O$_3$ glass-ceramics for self-cleaning applications. *J. Am. Ceram. Soc.* **2018**, *101*, 2901–2913. [CrossRef]
31. Zhao, J.; Pan, H.; He, X.; Wang, Y.; Gu, L.; Hu, Y.S.; Chen, L.; Liu, H.; Dai, S. Size-controlled synthesis and morphology evolution of bismuth trifluoride nanocrystals via a novel solvent extraction route. *Nanoscale* **2013**, *5*, 518–522. [CrossRef]
32. Feng, C.; Teng, F.; Liu, Z.; Chang, C.; Zhao, Y.; Wang, S.; Chen, M.; Yao, W.; Zhu, Y. A newly discovered BiF$_3$ photocatalyst with a high positive valence band. *J. Mol. Catal. A Chem.* **2015**, *401*, 35–40. [CrossRef]
33. Yang, Z.; Wang, X.; Liu, L.; Yang, S.; Su, X. First-principles calculations on structural, magnetic and electronic properties of oxygen doped BiF$_3$. *Comput. Mater. Sci.* **2011**, *50*, 3131–3135. [CrossRef]
34. Lei, S.; Wang, C.; Cheng, D.; Gao, X.; Chen, L.; Yan, Y.; Zhou, J.; Xiao, Y.; Cheng, B. Hierarchical BiF$_3$-Bi$_2$NbO$_5$F core-shell structure and its application in the photosensitized degradation of rhodamine B under visible light irradiation. *J. Phys. Chem. C* **2015**, *119*, 502–511. [CrossRef]
35. Sarkar, S.; Dash, A.; Mahalingam, V. Strong stokes and upconversion luminescence from ultrasmall Ln^{3+}-doped BiF$_3$ (Ln=Eu^{3+}, Yb^{3+}/Er^{3+}) nanoparticles confined in a polymer matrix. *Chem. Asian J.* **2014**, *9*, 447–451. [CrossRef] [PubMed]
36. Kan, Y.; Teng, F.; Yang, Y.; Xu, J.; Yang, L. Direct conversion mechanism from BiOCl nanosheets to BiOF, Bi$_7$F$_{11}$O$_5$ and BiF$_3$ in the presence of a fluorine resource. *RSC Adv.* **2016**, *6*, 63347–63357. [CrossRef]
37. Yu, J.; Li, C.; Liu, S. Effect of PSS on morphology and optical properties of ZnO. *J. Colloid Interface Sci.* **2008**, *326*, 433–438. [CrossRef] [PubMed]
38. Mills, A.; Wang, J.; McGrady, M. Method of rapid assessment of photocatalytic activities of self-cleaning films. *J. Phys. Chem. B* **2006**, *110*, 18324–18331. [CrossRef] [PubMed]
39. Lu, H.; Ju, T.; She, H.; Wang, L.; Wang, Q. Microwave-assisted synthesis and characterization of BiOI/BiF$_3$ p–n heterojunctions and its enhanced photocatalytic properties. *J. Mater. Sci. Mater. Electron.* **2020**, *31*, 13787–13795. [CrossRef]
40. Yang, Y.; Teng, F.; Kan, Y.; Yang, L.; Liu, Z.; Gu, W.; Zhang, A.; Hao, W.; Teng, Y. Investigation of the charges separation and transfer behavior of BiOCl/BiF$_3$ heterojunction. *Appl. Catal. B Environ.* **2017**, *205*, 412–420. [CrossRef]
41. Ritika; Kaur, M.; Umar, A.; Mehta, S.K.; Kansal, S.K. BiF$_3$ octahedrons: A potential natural solar light active photocatalyst for the degradation of Rhodamine B dye in aqueous phase. *Mater. Res. Bull.* **2019**, *112*, 376–383. [CrossRef]
42. Zhang, S.; Chen, X.; Song, L. Preparation of BiF$_3$/BiOBr heterojunctions from microwave-assisted method and photocatalytic performances. *J. Hazard. Mater.* **2019**, *367*, 304–315. [CrossRef] [PubMed]
43. Sharma, S.K.; Singh, V.P.; Chauhan, V.S.; Kushwaha, H.S.; Vaish, R. Photocatalytic Active Bismuth Fluoride/Oxyfluoride Surface Crystallized 2Bi$_2$O$_3$-B$_2$O$_3$ Glass–Ceramics. *J. Electron. Mater.* **2018**, *47*, 3490–3496. [CrossRef]
44. Singh, V.P.; Vaish, R. Controlled crystallization of photocatalytic active Bismuth oxyfluoride/Bismuth fluoride on SrO-Bi$_2$O$_3$-B$_2$O$_3$ transparent glass ceramic. *J. Eur. Ceram. Soc.* **2018**, *38*, 3635–3642. [CrossRef]
45. Singh, G.; Singh, V.P.; Vaish, R. Controlled crystallization of BiOCl/BiF$_3$ on ZnO–Bi$_2$O$_3$–B$_2$O$_3$ glass surfaces for photocatalytic and self-cleaning applications. *Materialia* **2019**, *5*, 100196. [CrossRef]

Article

Contribution of Oxide Supports in Nickel-Based Catalytic Elimination of Greenhouse Gases and Generation of Syngas

Wasim Ullah Khan [1,*], Mohammad Rizwan Khan [2], Rosa Busquets [3] and Naushad Ahmad [2]

1. Chemical and Process Engineering, College of Engineering, University of Canterbury, Christchurch 8041, New Zealand
2. Department of Chemistry, College of Science, King Saud University, P.O. Box. 2454, Riyadh 11451, Saudi Arabia; mrkhan@ksu.edu.sa (M.R.K.); anaushad@ksu.edu.sa (N.A.)
3. School of Life Sciences, Pharmacy and Chemistry, Kingston University London, Penrhyn Road, Kingston Upon Thames KT1 2EE, UK; r.busquets@kingston.ac.uk
* Correspondence: wasimkhan49@gmail.com

Abstract: Carbon dioxide and/or dry methane reforming serves as an effective pathway to mitigate these greenhouse gases. This work evaluates different oxide supports including alumina, Y-zeolite and H-ZSM-5 zeolite for the catalysis of dry reforming methane with Nickel (Ni). The composite catalysts were prepared by impregnating the supports with Ni (5%) and followed by calcination. The zeolite supported catalysts exhibited more reducibility and basicity compared to the alumina supported catalysts, this was assessed with temperature programmed reduction using hydrogen and desorption using carbon dioxide. The catalytic activity, in terms of CH_4 conversion, indicated that 5 wt% Ni supported on alumina exhibited higher CH_4 conversion (80.5%) than when supported on Y-zeolite (71.8%) or H-ZSM-5 (78.5%). In contrast, the H-ZSM-5 catalyst led to higher CO_2 conversion (87.3%) than Y-zeolite (68.4%) and alumina (83.9%) supported catalysts. The stability tests for 9 h time-on-stream showed that Ni supported with H-ZSM-5 had less deactivation (just 2%) due to carbon deposition. The characterization of spent catalysts using temperature programmed oxidation (O_2-TPO), X-ray diffraction (XRD) and thermo-gravimetric analysis (TGA) revealed that carbon deposition was a main cause of deactivation and that it occurred in the lowest degree on the Ni H-ZSM-5 catalyst.

Keywords: CO_2; CH_4; stability; H-ZSM-5; carbon deposition; greenhouse gas reduction

1. Introduction

Carbon dioxide (CO_2) and methane (CH_4) contribute to the increasing of the Earth's temperature. A reaction that can reduce the concentration of these gases is the dry reforming of methane, DRM (Equation (1)); this reaction can be very beneficial for the environment.

$$CH_4 + CO_2 \rightarrow 2CO + 2H_2, \Delta H° = 247.3 \text{ kJ mol}^{-1} \quad (1)$$

In addition to the possible use of DRM for mitigating greenhouse gases involved in global warming, it can generate sustainable hydrogen and syngas (CO + H_2). Syngas can be used as fuel and to produce a wide range of chemicals including methanol and hydrocarbons to make synthetic fuel [1]. DRM is of scientific and commercial importance [2,3].

Noble metal and base metal catalysts have been proposed for DRM [1]. Nickel-based metal catalysts have the potential to compete with noble metal-based catalysts when comparing their catalytic performance, cost, and abundance of the raw materials [2]. However, the existence of side reactions such as carbon monoxide disproportionation and methane decomposition brings challenge of coping with the deactivation of Ni catalysis due to coke deposits [2–4]. These are more problematic in DRM than in processes that include steam [4]. Moreover, sintering of metal particles also plays a role in decreasing the catalytic activity due to reducing the exposed active metal surface area. The improvement in catalytic activity performance and resistance to carbon deposition in Ni catalysis can be achieved by employing various preparation strategies leading to favorable structures (for instance, by adding promoters [5–9], selecting a suitable support which is usually a metal oxide [10–13], optimizing metal loading or using bimetallic active sites [3]). Among these strategies to reduce carbonaceous deactivation of the catalyst, catalyst supports offer advantages such as the possibility of dispersing metal catalyst nanoparticles over their surface. This enhances resistance to carbon deposition and can improve their catalytic activity [14]. The 3D network in zeolite supports, formed by shared oxygen atoms of SiO_4 and AlO_4 tetrahedra, possess high resistance to temperature and mechanical stability due to their ordered framework. Moreover, zeolites offer crystalline features such as uniform microporosity (diameter < 2 nm), and pore shape that gives selectivity by controlling the entrance and exit of reactant molecules in zeolite channels [15–18]. The unique shape selectivity feature of zeolites lies behind their application as catalysts and adsorbents in various industries [19,20]. Moreover, suitable properties of zeolites including their high specific surface areas, well-defined microporous structure, high thermal stability and high capacity of CO_2 adsorption [21,22] can play a role in catalysis. In addition, their potential in offering high metal dispersion, superior resistance to carbon formation, and suitable metal-support interaction has also attracted scientists towards utilizing zeolites as catalyst supports in the DRM reaction. Ni based catalysts supported on different zeolites have been reported for DRM reaction [23–29]. Hambali et al. [29] synthesized mesoporous fibrous MFI support via the microemulsion method and deposited Ni over support surfaces using double solvent, wetness impregnation, and physical mixing methods. The catalytic activity results revealed that Ni based catalysts prepared by wetness impregnation exhibited stable performance with least carbon deposition as compared with catalysts synthesized by double solvent or physical mixing approaches. It was found that the acidity of the support hindered the side reaction i.e., methane cracking mainly responsible for carbon deposition at high reaction temperatures (800 °C).

Keeping in mind the distinct properties of zeolites, this work fills the literature gap by providing an insight into the role of reducibility, carbon dioxide adsorption capacity, and basicity of zeolites in displaying stable performance during DRM. Furthermore, the aim of this research is to identify a conventional support that can enhance DRM, determine how zeolite supports can affect the DRM reaction, and determine how their stability and performance compares with conventional alumina supported Ni catalysts.

2. Materials and Methods

2.1. Catalyst Synthesis

In order to synthesize 5 wt% nickel-based catalysts, nitrate salt precursor [$(Ni(NO_3)_2 \cdot 6H_2O)$] was wet impregnated over different supports including Y-zeolite (Y), H-ZSM-5 (H) and alumina (A) to synthesize catalysts denoted as NY, NH, and NA, where N represents 5 wt% Ni and Y, and H and A represent respective supports. Briefly, 30 mL of 28 mM of nickel precursor solution were added dropwise to a suspension of the individual oxide supports (33 g/L) in deionized water stirred with magnetic bar at 300 rpm. This process was carried out at 60 °C. The catalysts were subsequently dried overnight (12 h) at 110 °C under atmospheric before being subjected to calcination at 500 °C in MTI®® furnace for 3 h under air atmosphere. The catalyst preparation steps and DRM over catalyst surface are pictorially depicted in Scheme 1.

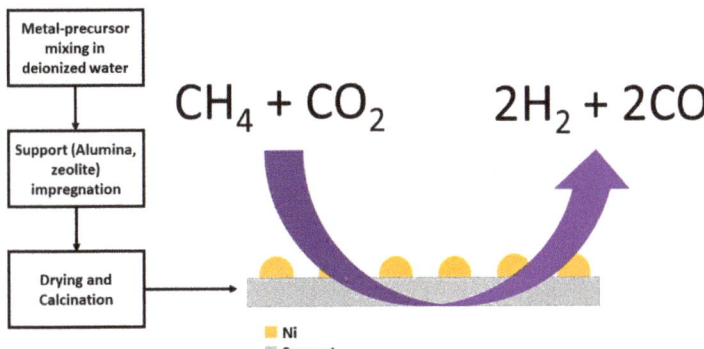

Scheme 1. Catalyst preparation steps and depiction of reforming reaction over catalyst.

2.2. Catalytic Testing

Catalytic activity tests for dry reforming reaction were carried out in a 10 mm i.d. and 40 cm long stainless steel-tube at atmospheric pressure using 600 mg of a catalyst sample. Prior to the reaction, hydrogen gas (40 mL/min) was used to activate the catalyst and to reduce the nickel oxide at 550 °C for 90 min. Nitrogen was flown through the reactor to purge hydrogen and to cool down the reactor to the target reaction temperature. The feed gas had volume ratio of 17/17/2 corresponding to methane/carbon dioxide/nitrogen mixture with total feed flow rate of 36 mL/min. The products and unconverted reactants were monitored by an online gas chromatograph (SRI Instruments) equipped with a flame ionization detector (FID) and a thermal conductivity detector (TCD).

2.3. Catalyst Characterization

BET specific surface area measurements were conducted on a Micromeritics Physisorption Unit (Gemini VI) by using a 300 mg sample. All of the samples were pre-treated before analysis to remove any impurities and subsequently analyzed under a liquid nitrogen atmosphere. X-ray diffraction (XRD) analysis of powder samples were carried out using a Rigaku (Miniflex) diffractometer equipped with radiation source of Cu Kα radiation which was operated at 40 kV and 40 mA. Temperature-programmed characterizations including desorption (TPD), reduction (TPR), and oxidation (TPO) were conducted using chemisorption apparatus (Micromeritics AutoChem II apparatus). For H_2-TPR, pretreatment of 50 mg of the catalyst sample was carried out under Argon (Ar) flowing at 20 mL/min at 150 °C for 30 min. Then, the sample was cooled down to room temperature before the temperature was raised to 1000 °C using a furnace and temperature ramp rate of 10 °C/min under mixture of 10%H_2 in Ar flowing at 40 mL/min. For CO_2-TPD measurements, same amount of sample (50 mg) was pretreated at 200 °C under helium (He) flowing at 20 mL/min for 1 h followed by cooling the sample down to 50 °C at which CO_2 was adsorbed for 30 min by flowing mixture of 10%CO_2 in helium at 30 m/min. The desorption profiles were recorded using thermal conductivity detector (TCD) by raising temperature linearly from 60 to 800 °C using 10 °C/min. Carbon deposition and type of graphitic carbon were analyzed using TPO experiments in which catalyst samples were pretreated at 150 °C for 30 min under helium flowing at 30 mL/min followed by cooling the samples down to room temperature. The oxidation profiles were measured by raising the sample temperature to 1000 °C using 10 °C/min under mixture of 10%O_2 in helium flowing at 30 mL/min. The weight loss of the spent catalysts samples was analyzed using the thermogravimetric analysis (TGA) where 20 mg of each sample was subjected to heating under air from room temperature to 800 °C using 10 °C/min.

3. Results

3.1. Catalytic Activity

5 wt% Ni supported over γ-Al_2O_3, Y-zeolite and H-ZSM-5 catalysts (for simplification, these catalysts are designated as NA, NY and NH, respectively) were studied for DRM reaction at different temperatures from 500 to 700 °C. Figure 1a,b shows activity performance in terms of CH_4 conversion and H_2/CO ratios versus temperature for the NA, NY, and NH catalysts, respectively. The activity results demonstrate an increase in CH_4 conversion with increase in reaction temperature from 500 to 700 °C which confirms the endothermic nature of DRM reaction [30,31]. The ratios of H_2 to CO show interesting trends for each catalyst. It is noteworthy that CH_4 conversion is mainly responsible for H_2 production while CO comes from CO_2. Hence, H_2/CO ratios less than unity clearly justifies CO_2 conversions are higher than CH_4 conversion and vice versa. CO_2 conversions higher than CH_4 conversions also proves the occurrence of reverse water gas shift reaction ($CO_2 + H_2 \leftrightarrow CO + H_2O$) which consumes H_2 and brings H_2/CO ratios less than one. This also suggests that H_2/CO ratios higher than one, resulting from higher CH_4 conversions, lead to the catalysts being more prone to carbon deposition [31].

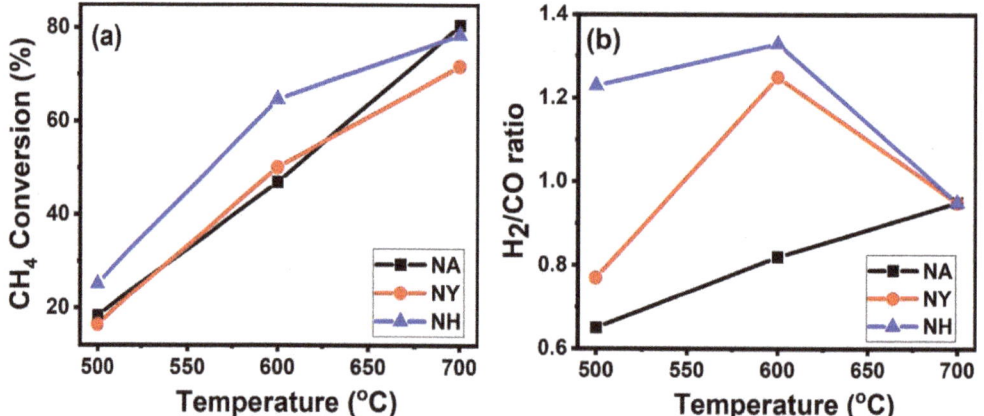

Figure 1. (a) CH_4 and (b) CO_2 conversion versus time on stream for Ni supported on alumina (NA), Ni supported on Y-zeolite (NY), and Ni supported on H-ZSM-5 (NH) catalysts.

3.2. Catalytic Stability

The deposition of carbon over the surface of a catalyst is crucial to investigate in examining DRM reactions. The stability of a catalyst is evaluated by testing its activity for a prolonged time to estimate the loss in activity. Figure 2 presents CH_4 and CO_2 conversions as a function of time for 9 h time on stream at fixed reaction temperature of 700 °C. The decrease in both CH_4 and CO_2 conversions was obvious for all of the tested catalysts (NA, NY and NH) and it was associated with the deactivation of all the catalysts. Interestingly despite that all catalysts showed deactivation, the extent of deactivation was not the same for all the catalysts. From quantitative results in Table 1, the NH catalyst showed negligible loss in CH_4 and CO_2 conversions as compared with NA (2.3 and 5.2% respectively) and NY (5 and 1.1% respectively) catalysts. The deactivation factor as function of extent of deactivation or catalyst stability was found to be the lowest (2%) in the NH catalyst.

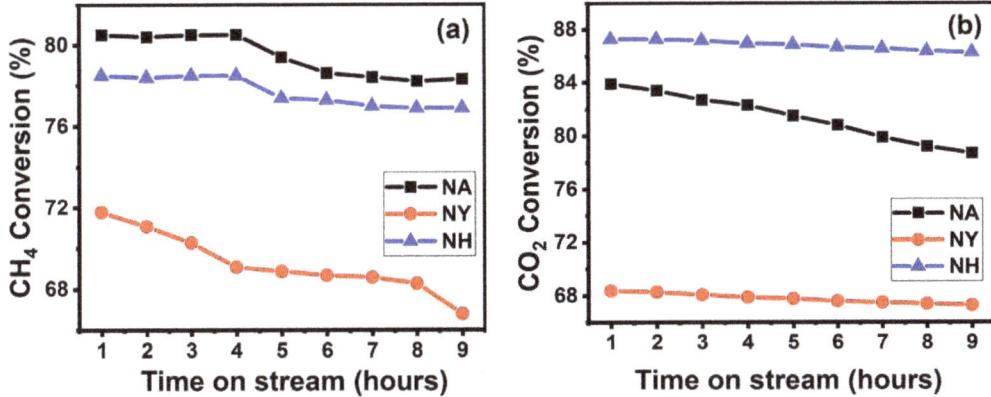

Figure 2. (a) CH$_4$ and (b) CO$_2$ conversion versus time on stream for Ni supported on alumina (NA), Ni supported on Y-zeolite (NY), and Ni supported on H-ZSM-5 (NH) catalysts.

Table 1. Performance of NA, NY, and NH catalysts in CO$_2$ reforming of methane after 9 h.

Catalyst	% DF [a]	Coke (wt. %) [b]	S$_{BET}$ (m^2/g) [c]	S$_{BET}$ (m^2/g) [d]
NH	2.0	3.7	335.3	318.4
NY	6.9	14.7	573.3	498.1
NA	2.9	8.5	209.7	171.7

[a] Deactivation Factor (%DF) = [([CH$_4$]$_{initial}$ − [CH$_4$]$_{final}$)/[CH$_4$]$_{initial}$] × 100; [b] Determined by TGA; [c] Before reaction; [d] After reaction.

Thermodynamically two of the side reactions including methane decomposition and Boudouard reaction (Equation (2)) are mainly forming carbon over the catalyst surface under reaction conditions [30].

$$2CO \rightleftharpoons CO_2 + C \qquad (2)$$

In order to investigate whether carbon formation was the main source of deactivation, thermogravimetric analysis (TGA) was used to quantify the carbon deposition. It can be clearly seen from TGA results in Table 1 that the NH catalyst exhibited weight loss of 3.7 wt% much lesser than NY (14.7 wt%) and NA (8.5 wt%). Hence, it can be concluded that the NH catalyst proves to show long term stability and lowest amount of carbon deposition. These findings are further discussed based on the characterizations of the catalysts before and after reaction as described in the following paragraphs.

The stability of the best catalyst (NH) in the current study is compared with the catalysts supported on zeolite or oxide supports (Table 2) and deactivation factor was used as a measure of stability. The deactivation factor over 5%Ni-ZSM [32] catalyst for similar reaction duration of 9 h was much higher (24.2%) than the NH catalyst (2%). Even Ni and Co based bimetallic catalyst supported on ZSM5 (1Co2Ni-ZSM5) [28] showed deactivation factor over 6 times higher (13.6%) than the NH catalyst. These catalysts were even tested at higher temperature (800 °C) as compared with the current NH catalyst (tested at 700 °C). Comparing the NH catalyst with alumina (2.25%Sr-10%Co/Al$_2$O$_3$) [33], ceria (Ni-Ce Imp) [34], and MCM41 [35] supported catalysts, the later catalysts showed more deactivation even with shorter reaction duration. Hence it can be concluded that the NH catalyst presents potentially more stable performance than other zeolite supported catalysts as well as catalysts with higher metal contents or employed promoters.

Table 2. Comparison of current work with the previously reported work.

Catalyst	Reaction Temp. (°C)	Initial CH$_4$ Conversion (%)	%DF	TOS (h)	Ref.
7% Ni/ZSM-5	700	91	35.3	5	[14]
7% Ni/ Zeolite Y	700	92	0.43	5	[14]
1Co2Ni-ZSM5	800	66	13.6	12	[28]
5%Ni-ZSM	800	96.2	24.2	9	[32]
2.25%Sr-10%Co/Al$_2$O$_3$	700	80.1	2.5	6	[33]
Ni-Ce Imp	700	81.1	3.7	6	[34]
5%Ni + 1%Sc/MCM41	800	65	19.2	~7	[35]
NH	700	78.5	2.0	9	This work

3.3. Catalysts Characterization

The specific surface areas measured by Brunauer-Emmett-Teller (BET) isotherms of all of the catalysts before and after DRM reaction are given in Table 1. The NY catalyst is found to have highest specific surface area (573.3 m^2/g) before reaction but NY catalyst also showed significant loss (13%) in specific surface area during reaction. The NA catalyst showed a loss of 18% in specific surface area while NH catalyst exhibited a minimum loss of 5%. The loss in specific surface area is accounted for carbon formation over the catalyst surface during DRM reaction. The minimal loss in specific surface area for NH catalyst indicates its superior stability under reaction conditions.

In order to estimate the crystallinity and different phases of the as-synthesized catalysts, X-ray diffraction (XRD) study was conducted over NA, NY and NH fresh catalysts and NA used catalyst. From Figure 3, the characteristic peaks of NiO were found over all the catalysts at two theta values of 44 and 63° (JCPDS: 01-073-1519), except for the fact that NY and NH catalysts showed additional peak of NiO at 37° (JCPDS: 01-073-1519). The peaks at 42 and 66° in the case of NA catalysts are characteristics of alumina support (JCPDS: 00-004-0875). The diffraction peaks had higher intensities in case of NY and NH catalysts as compared with NA catalyst showing higher crystallinity of the former catalysts. The NA catalyst after reaction revealed the formation of carbon during reaction as indicated by strong diffraction peak at 26° associated with graphite. Moreover, in addition to Ni^0, the NA catalyst after reaction also showed a characteristic peak of spinel NiAl$_2$O$_4$ (JCPDS: 01-073-0239) at 45 and 78°.

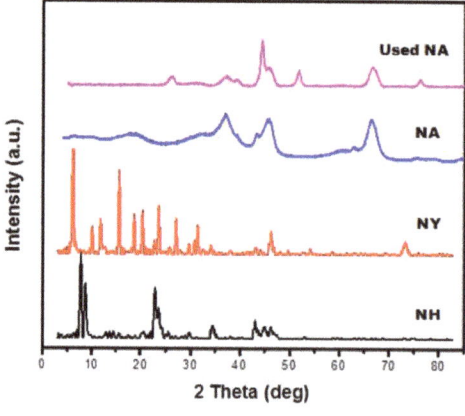

Figure 3. XRD patterns for Ni supported on alumina (NA), Ni supported on Y-zeolite (NY), and Ni supported on H-ZSM-5 (NH) fresh and Ni supported on alumina (NA) used catalysts.

The temperature-programmed oxidation (TPO) is a useful technique to estimate the amount of carbon formed during reaction by oxidizing the carbon as function of temperature. TPO also provides significant details about the structure, morphology and composition of accumulated carbon. Figure 4 displays the TPO profiles as a function of temperature for all the three catalysts (NA, NY and NH respectively). The spent or used catalysts exhibited carbon gasification peaks in the temperature range from 400 to 640 °C associated with different types of carbon including carbon nanofibers and carbon nanotubes.

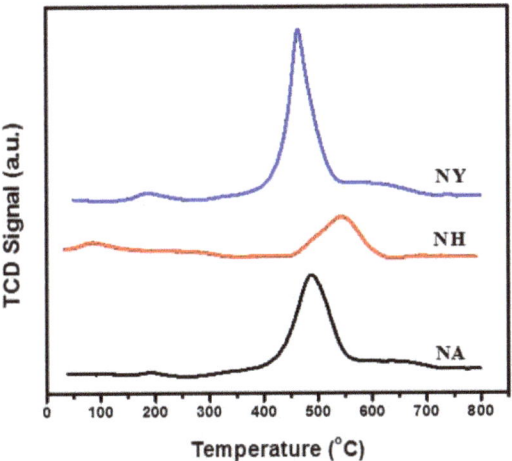

Figure 4. TPO profiles for Ni supported on alumina (NA), Ni supported on Y-zeolite (NY), and Ni supported on H-ZSM-5 (NH) spent catalysts.

The following four side reactions are most probably responsible for carbon deposition during DRM reaction.

$$2CO \leftrightarrow CO_2 + C \tag{3}$$

$$CH_4 \leftrightarrow 2H_2 + C \tag{4}$$

$$CO + H_2 \leftrightarrow H_2O + C \tag{5}$$

$$CO_2 + 2H_2 \leftrightarrow 2H_2O + C \tag{6}$$

As mentioned earlier, the first two reactions are favorable at high temperatures while the last two reactions (reactions (5) and (6)) require lower temperatures [36]. The peak temperature indicates the type of carbon formed, and hence it is obvious that the carbon formed over the surface of NY (corresponding peak temperature of ~460 °C) and NY (corresponding peak temperature of ~490 °C) is less crystalline and easily gasified as compared with crystalline carbon deposited over the surface of NH catalyst (corresponding peak temperature of ~545 °C). The peak intensity shows the amount of carbon formed, and it is obvious that NH catalyst had less amount of carbon deposition among all of the catalysts [37].

The extent of interaction between the metal and support plays a vital role in catalytic activity. Temperature-programmed reduction (TPR) using hydrogen is utilized to measure the metal-support interaction for NA, NY, and NH catalysts and extent of reducibility of each catalyst. Figure 5 shows the reduction profiles for NA, NY, and NH fresh catalysts as a function of temperature. NY catalyst exhibited two peaks at 350 and 440 °C along with a shoulder centered at 550 °C. The two low temperature peaks are assigned to the reduction of NiO species weakly interacting with the support while shoulder is ascribed to the reduction of NiO species having medium interaction with the support (Y-zeolite). Also, the NY catalyst shows NiO species which are easier to reduce as compared with the NA and NH catalysts.

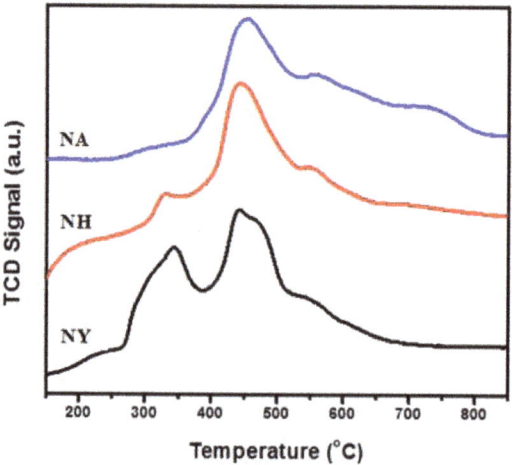

Figure 5. TPR profiles for Ni supported on alumina (NA), Ni supported on Y-zeolite (NY), and Ni supported on H-ZSM-5 (NH) fresh catalysts.

In case of NH catalyst, the first smaller peak is centered at 330 °C and a second large peak has maxima at 450 °C while the shoulder is centered at 560 °C. This suggests that the strength of NiO species over NH catalyst is not significantly different than the NY catalyst. On the contrary, NA catalyst gave only one distinct reduction peak at 455 °C with two smaller shoulders centered at 560 and 740 °C, respectively. It is interesting to note that lower temperature peak appearing in the 330–350 °C range is disappeared in NA catalyst while a new shoulder at 740 °C is observed. These results suggest that NA catalyst showed medium to strong metal-support interactions. The shoulder at 740 °C is assigned to the reduction of spinel $NiAl_2O_4$ species.

The extent of basic sites and their strength for fresh NA, NY, and NH catalysts was measured by employing temperature-programmed desorption (TPD) using CO_2. The basicity of the catalyst is found to influence the carbon deposition and a higher number of basic sites led to lesser amount of carbon deposition [38]. The enhanced basicity promotes CO_2 activation over the surface of the catalyst which reacts with carbon formed due to side reactions. Consequently, reverse Boudouard reaction ($2CO \rightleftharpoons CO_2 + C$) converts the carbonaceous species into CO. Hence, the catalyst with higher basicity is expected to show the minimum carbon deposition. CO_2-TPD profiles are shown in Figure 6 for fresh NA, NY, and NH catalysts.

NA catalyst presents two peaks, centered at 80 and 260 °C, along with a shoulder centered at 560 °C. The two peaks which appeared at lower temperatures are associated with weak basic sites and the shoulder represents strong basic sites. In the case of the NY catalyst, a small peak appears at 150 °C while a broad peak is centered at 275 °C and both are assigned to weak basic sites. NH catalyst shows two peaks with peak maxima at 130 and 300 °C respectively along with a shoulder centered at 550 °C. The first peak represents weak basic sites and the second peak is associated with medium basic sites while the shoulder is assigned to strong basic sites. It is noteworthy that the peaks for NH catalysts are broader than NY and NA catalysts and hence exhibit more amount of CO_2 adsorbed which leads to lesser carbon formation. This is in agreement with the TGA and TPO results.

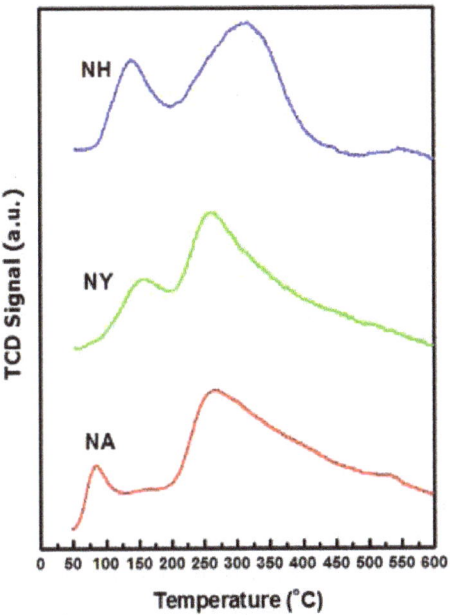

Figure 6. CO$_2$-TPD for Ni supported on alumina (NA), Ni supported on Y-zeolite (NY), and Ni supported on H-ZSM-5 (NH) fresh catalysts.

4. Discussion

Prior to DRM reaction study, as-synthesized Ni based catalysts were characterized to assess their potential performance in the reforming reaction. High specific surface areas in zeolites are known to influence the catalytic performance during DRM [39]. It is noteworthy that despite the higher specific surface area of the study Y-zeolite, as compared with H-ZSM-5 and alumina, the Y-zeolite supported Ni catalyst (NY) exhibited less conversions at lower reaction temperatures (Figure 1). This could be attributed to the zeolite having pores filled with nickel particles and/or the presence of nickel particles entrapped in the pores of zeolite and hence these were not available for DRM reaction [40]. Furthermore, the loss of specific surface area (Table 1) in the spent catalysts also demonstrates the drastic role of carbon deposition during DRM. The XRD diffraction profiles from the as-synthesized catalysts (Figure 3) have helped to identify crystalline phases (nickel oxide, alumina, and zeolites) in the composites. The presence of spinel species such as nickel aluminate was also obvious. Moreover, the deposition of crystalline graphitic carbon was further confirmed by XRD data of spent NA catalyst. The reduction profiles using TPR (Figure 5) demonstrate easier reduction of Ni oxides or higher reducibility of the Y-zeolite supported catalysts (NY) as compared to the alumina (NA) and H-ZSM-5 supported catalysts (NH). The basicity of the composites, assessed in terms of temperature programmed desorption data using carbon dioxide as probe gas, showed that the H-ZSM-5 supported catalyst (NH) exhibited larger CO$_2$ adsorption capacity than the rest of the study catalysts. The pre-reforming characterization results indicated larger specific surface area, higher number of reducible species and relatively lower basicity in the Y-zeolite supported catalyst (NY). These characteristics lead to expect better performance by NY during the reforming reaction however the activity results (Figures 1 and 2) did not follow the prediction. The DRM mechanism requires adsorption of reactants over the active sites of the catalyst in its initial step. Subsequently, they will react leading to products that ultimately will leave the catalyst surface following desorption [30]. CH$_4$ adsorption requires Ni in its metallic form as the active site [40].

The catalytic stability (Figure 2) in terms of both CH_4 and CO_2 conversions showed loss of activity over time for all catalysts with varying deactivation degrees. The decrease in initial conversions for zeolite supported catalysts can be assigned to loss of Ni active sites residing inside the pores in the zeolite. The CO_2 adsorption capacity is also found to influence the catalytic performance and hence more CO_2 conversion is evidenced from H-ZSM-5 supported catalyst (NH) as compared to other catalysts in agreement with CO_2-TPD results (Figure 6). Higher CO_2 conversion helps to provide an oxidative environment to gasify carbon deposited over the surface of the catalyst and hence this can lead to less deactivation. Thus, the catalyst with the highest CO_2 adsorption capacity and CO_2 conversion i.e., H-ZSM-5 supported catalyst (NH), has shown to present the lowest deactivation factor: just 2%. The main cause of deactivation was found to be carbon deposition according to TPO (Figure 4) and TGA (Table 1) results. The catalyst system studied in this work demonstrates the role of catalyst's properties such as reduction behavior, CO_2 adsorption capacity and basicity of commercial zeolites in comparison with conventional alumina supported catalysts during DRM. A comprehensive study of the surface chemistry of the hierarchical zeolites using spectroscopic techniques and their performance evaluation in relation to commercial zeolites is planned.

5. Conclusions

The elimination of greenhouse gases via reforming CH_4 with CO_2, was investigated via heterogeneous catalysis. Specifically, the role of different supports (alumina and zeolites Y-zeolite and H-ZSM-5) for Ni-based catalysts were investigated for DRM. The activity results showed that Ni deposited onto H-ZSM-5 (NH catalyst) exhibited excellent stability (just 2% deactivation over 9h on-stream) with the least amount of carbon deposition when compared with the Ni deposited onto alumina and Y-zeolite supports. The least amount of carbon deposition was confirmed with TGA and TPO. The favorable metal-support interaction and high basicity of Ni catalytic sites onto the H-ZSM-5 contributed towards its stable performance. A comparison between the prepared NH catalyst with already reported catalysts for DRM reaction showed that the NH catalyst developed here, with 87.3% conversion of CO_2, outperformed other Ni based catalysts and even bimetallic catalysts. The catalyst where Ni was deposited onto alumina showed the best conversion of CH_4 (80.5%) compared to the zeolite supports, which led to up to 78.5% CH_4 conversion. Overall, Ni deposited onto H-ZSM-5 constitutes a catalyst with exceptional performance for the reduction of the greenhouse gases CO_2 and CH_4 (87.3 and 78.5% conversion, respectively) and generation of syngas. This study provides a platform to further evaluate the performance of hierarchical zeolites synthesized by modifying conventional zeolites.

Author Contributions: Conceptualization, W.U.K.; methodology, M.R.K. and N.A.; formal analysis, W.U.K.; writing—original draft preparation, W.U.K.; writing—review and editing, M.R.K., R.B. and N.A.; funding acquisition, M.R.K. All authors have read and agreed to the published version of the manuscript.

Funding: The APC was funded by Researchers Supporting Project No. (RSP-2021/138) King Saud University, Riyadh, Saudi Arabia.

Institutional Review Board Statement: Not applicable.

Informed Consent Statement: Not applicable.

Data Availability Statement: Not applicable.

Acknowledgments: The authors would like to thank the Researchers Supporting Project No. (RSP-2021/138) King Saud University, Riyadh, Saudi Arabia for funding this project.

Conflicts of Interest: The authors declare no conflict of interest.

References

1. Arora, S.; Prasad, R. An overview of dry reforming of methane: Strategies to reduce carbonaceous deactivation of catalysts. *RSC Adv.* **2016**, *6*, 108668–108688. [CrossRef]
2. Li, M.; Sun, Z.; Hu, Y.H. Catalysts for CO_2 reforming of CH_4: A review. *J. Mater. Chem. A* **2021**, *9*, 12495–12520. [CrossRef]
3. Bao, Z.; Yu, F. Catalytic Conversion of Biogas to Syngas via Dry Reforming Process. In *Advances in Bioenergy*; Li, Y., Ge, S., Eds.; Elsevier: Amsterdam, The Netherlands, 2018; Volume 3, pp. 43–76.
4. Wang, X.; Economides, M. *Advanced Natural Gas Engineering*, 1st ed.; Elsevier: Amsterdam, The Netherlands, 2009; pp. 1–368.
5. Daza, C.E.; Gallego, J.; Mondragon, F.; Moreno, S.; Molina, R. High stability of Ce-promoted Ni/Mg–Al catalysts derived from hydrotalcites in dry reforming of methane. *Fuel* **2010**, *89*, 592–603. [CrossRef]
6. Jeong, D.W.; Jang, W.J.; Shim, J.O.; Roh, H.S.; Son, I.H.; Lee, S.J. The effect of preparation method on the catalytic performance over superior MgO-promoted Ni–$Ce_{0.8}Zr_{0.2}O_2$ catalyst for CO_2 reforming of CH_4. *Int. J. Hydrogen Energy* **2013**, *38*, 13649–13654. [CrossRef]
7. Son, I.H.; Lee, S.J.; Song, I.Y.; Jeon, W.S.; Jung, I.; Yun, D.J.; Jeong, D.W.; Shim, J.O.; Jang, W.J.; Roh, H.S. Study on coke formation over Ni/γ-Al_2O_3, Co-Ni/γ-Al_2O_3, and Mg-Co-Ni/γ-Al_2O_3 catalysts for carbon dioxide reforming of methane. *Fuel* **2014**, *136*, 194–200. [CrossRef]
8. Quiroga, M.M.B.; Luna, A.E.C. Catalytic activity and effect of modifiers on Ni-based catalysts for the dry reforming of methane. *Int. J. Hydrogen Energy* **2010**, *35*, 6052–6056. [CrossRef]
9. Khan, A.; Sukonket, T.; Saha, B.; Idem, R. Catalytic activity of various 5 wt.% Ni/$Ce_{0.5}Zr_{0.33}M_{0.17}O_{2-\delta}$ catalysts for the CO_2 reforming of CH_4 in the presence and absence of steam. *Energy Fuels* **2012**, *26*, 365–379. [CrossRef]
10. Chen, H.W.; Wang, C.Y.; Yu, C.H.; Tseng, L.T.; Liao, P.H. Carbon dioxide reforming of methane reaction catalyzed by stable nickel copper catalysts. *Catal. Today* **2004**, *97*, 173–180. [CrossRef]
11. Pompeo, F.; Nichio, N.N.; Ferretti, O.A.; Resasco, D. Study of Ni catalysts on different supports to obtain synthesis gas. *Int. J. Hydrogen Energy* **2005**, *30*, 1399–1405. [CrossRef]
12. Damyanova, S.; Pawelec, B.; Arishtirova, K.; Fierro, J.L.G. Ni-based catalysts for reforming of methane with CO_2. *Int. J. Hydrogen Energy* **2012**, *37*, 15966–15975. [CrossRef]
13. Hu, Y.H.; Ruckenstein, E. Catalytic conversion of methane to synthesis gas by partial oxidation and CO_2 reforming. *Adv. Catal.* **2004**, *48*, 297–345.
14. Luengnaruemitchai, A.; Kaengsilalai, A. Activity of different zeolite-supported Ni catalysts for methane reforming with carbon dioxide. *Chem. Eng. J.* **2008**, *144*, 96–102. [CrossRef]
15. Flanigen, E.M.; Jansen, J.; Bekkum, H.V. *Introduction to Zeolite Science and Practice*, 2nd ed.; Elsevier: Amsterdam, The Netherlands, 1991; pp. 1–1078.
16. Csicsery, S.M. Shape-selective catalysis in zeolites. *Zeolites* **1984**, *4*, 202–213. [CrossRef]
17. Khouw, C.B.; Dartt, C.B.; Labinger, J.A.; Davis, M.E. Studies on the Catalytic-Oxidation of Alkanes and Alkenes by Titanium Silicates. *J. Catal.* **1994**, *149*, 195–205. [CrossRef]
18. Kareem, A.; Chand, S.; Mishra, I. Disproportionation of Toluene to Produce Benzene and p-Xylene: A Review. *J. Sci. Ind. Res.* **2001**, *60*, 319–327.
19. Marcilly, C.R. Where and how shape selectivity of molecular sieves operates in refining and petrochemistry catalytic processes. *Top. Catal.* **2000**, *13*, 357–366. [CrossRef]
20. Weitkamp, J. Zeolites and catalysis. *Solid State Ion.* **2000**, *131*, 175–188. [CrossRef]
21. Nimwattanakul, W.; Luengnaruemitchai, A.; Jitkarnka, S. Potential of Ni supported on clinoptilolite catalysts for carbon dioxide reforming of methane. *Int. J. Hydrogen Energy* **2006**, *31*, 93–100. [CrossRef]
22. Wang, K.; Li, X.; Ji, S.; Sun, S.; Ding, D.; Li, C. CO_2 reforming of methane to syngas over Ni/SBA-15/FeCrAl. *Stud. Surf. Sci. Catal.* **2007**, *167*, 367–372.
23. Wei, B.; Yang, H.; Hu, H.; Wang, D.; Jin, L. Enhanced production of light tar from integrated process of in-situ catalytic upgrading lignite tar and methane dry reforming over Ni/mesoporous Y. *Fuel* **2020**, *279*, 118533. [CrossRef]
24. Albarazi, A.; Beaunier, P.; Da Costa, P. Hydrogen and syngas production by methane dry reforming on SBA-15 supported nickel catalysts: On the effect of promotion by $Ce_{0.75}ZrO_{0.25}O_2$ mixed oxide. *Int. J. Hydrogen Energy* **2013**, *38*, 127–139. [CrossRef]
25. Vafaeian, Y.; Haghighi, M.; Aghamohammadi, S. Ultrasound assisted dispersion of different amount of Ni over ZSM-5 used as nanostructured catalyst for hydrogen production via CO_2 reforming of methane. *Energy Convers. Manag.* **2013**, *76*, 1093–1103. [CrossRef]
26. Bawah, A.-R.; Malaibari, Z.O.; Muraza, O. Syngas production from CO_2 reforming of methane over Ni supported on hierarchical silicalite-1 fabricated by microwave-assisted hydrothermal synthesis. *Int. J. Hydrogen Energy* **2018**, *43*, 13177–13189. [CrossRef]
27. Dai, C.; Zhang, S.; Zhang, A.; Song, C.; Shi, C.; Guo, X. Hollow zeolite encapsulated Ni–Pt bimetals for sintering and coking resistant dry reforming of methane. *J. Mater. Chem.* **2015**, *3*, 16461–16468. [CrossRef]
28. Estephane, J.; Aouad, S.; Hany, S.; Khoury, B.E.; Gennequin, C.; Zakhem, H.E.; Nakat, J.E.; Aboukais, A.; Aad, E.A. CO_2 reforming of methane over Ni–Co/ZSM5 catalysts. Aging and carbon deposition study. *Int. J. Hydrogen Energy* **2015**, *40*, 9201–9208. [CrossRef]
29. Hambali, H.U.; Jalil, A.A.; Abdulrasheed, A.A.; Siang, T.J.; Vo, D.V.N. Enhanced dry reforming of methane over mesostructured fibrous Ni/MFI zeolite: Influence of preparation methods. *J. Energy Inst.* **2020**, *93*, 1535–1543. [CrossRef]

30. Ibrahim, A.A.; Fakeeha, A.H.; Fatesh, A.S.A. Enhancing hydrogen production by dry reforming process with strontium promoter. *Int. J. Hydrogen Energy* **2014**, *39*, 1680–1687. [CrossRef]
31. Fatesh, A.S.A.; Ibrahim, A.A.; Fakeeha, A.H.; Soliman, M.A.; Siddiqui, M.R.; Abasaeed, A.E. Coke formation during CO_2 reforming of CH_4 over alumina-supported nickel catalysts. *Appl. Catal. A Gen.* **2009**, *364*, 150–155. [CrossRef]
32. Sarkar, B.; Tiwari, R.; Singha, R.K.; Suman, S.; Ghosh, S.; Acharyya, S.S.; Mantri, K.; Konathala, L.N.S.; Pendem, C.; Bal, R. Reforming of methane with CO_2 over Ni nanoparticle supported on mesoporous ZSM-5. *Catal. Today* **2012**, *198*, 209–214. [CrossRef]
33. Fakeeha, A.H.; Naeem, M.A.; Khan, W.U.; Fatesh, A.S.A. Syngas production via CO_2 reforming of methane using Co-Sr-Al catalyst. *J. Ind. Eng. Chem.* **2014**, *20*, 549–557. [CrossRef]
34. Naeem, M.A.; Fatesh, A.S.A.; Abasaeed, A.E.; Fakeeha, A.H. Activities of Ni-based nano catalysts for CO_2–CH_4 reforming prepared by polyol process. *Fuel Process. Technol.* **2014**, *122*, 141–152. [CrossRef]
35. Fatesh, A.S.A.; Atia, H.; Dahrieh, J.K.A.; Ibrahim, A.A.; Eckelt, R.; Armbruster, U.; Abasaeed, A.E.; Fakeeha, A.H. Hydrogen production from CH_4 dry reforming over Sc promoted Ni/MCM-41. *Int. J. Hydrogen Energy* **2019**, *44*, 20770–20781. [CrossRef]
36. Laosiripojana, N.; Sutthisripok, W.; Assabumrungrat, S. Synthesis gas production from dry reforming of methane over CeO_2 doped Ni/Al_2O_3: Influence of the doping ceria on the resistance toward carbon formation. *Chem. Eng. J.* **2005**, *112*, 13–22. [CrossRef]
37. de Sousa, F.F.D.; de Sousa, H.S.A.; Oliveira, A.C.; Junior, M.C.C.; Ayala, A.P.; Barros, E.B.; Viana, B.C.; Filho, J.M.; Oliveira, A.C. Nanostructured Ni-containing spinel oxides for the dry reforming of methane: Effect of the presence of cobalt and nickel on the deactivation behaviour of catalysts. *Int. J. Hydrogen Energy* **2012**, *37*, 3201–3212. [CrossRef]
38. Moradi, G.; Khezeli, F.; Hemmati, H. Syngas production with dry reforming of methane over Ni/ZSM-5 catalysts. *J. Nat. Gas Sci. Eng.* **2016**, *33*, 657–665. [CrossRef]
39. Najfach, A.J.; Almquist, C.B.; Edelmann, R.E. Effect of Manganese and zeolite composition on zeolite-supported Ni-catalysts for dry reforming of methane. *Catal. Today* **2021**, *369*, 41–47. [CrossRef]
40. Al-Fatesh, A.S.; Arafat, Y.; Atia, H.; Ibrahim, A.A.; Ha, Q.L.M.; Schneider, M.; M-Pohl, M.; Fakeeha, A.H. CO_2-reforming of methane to produce syngas over Co-Ni/SBA-15 catalyst: Effect of support modifiers (Mg, La and Sc) on catalytic stability. *J. CO_2 Util.* **2017**, *21*, 395–404. [CrossRef]

Article

High Dispersion of CeO$_2$ on CeO$_2$/MgO Prepared under Dry Conditions and Its Improved Redox Properties

Kenji Taira * and Reiko Murao

Advanced Technology Research Laboratories, Nippon Steel Corporation, 20-1 Shintomi, Futtsu 293-8511, Chiba, Japan
* Correspondence: taira.e84.kenji@jp.nipponsteel.com; Tel.: +81-70-3914-4689

Abstract: Suppressing the usage of rare-earth elements is crucial for making the catalysts sustainable. Preparing CeO$_2$ nanoparticles is a common technique to reduce CeO$_2$ consumption, but such nanoparticles are prone to sinter or react with the supports when subjected to heat treatments. This study demonstrated that stable CeO$_2$ nanoparticles were deposited on MgO by the simple impregnation method. When CeO$_2$/MgO was prepared under the dry atmosphere, the CeO$_2$ nanoparticles remained ~3 nm in diameter even after being heated at 800 °C, which is much smaller than ~5 nm of CeO$_2$/MgO prepared under ambient air. Temperature-programmed reduction, temperature-programmed oxidation, X-ray photoelectron spectroscopy, and in situ X-ray diffraction studies showed that CeO$_2$/MgO exhibited higher oxygen mobility when prepared under the dry atmosphere. Dry reforming reaction demonstrated that CeO$_2$/MgO prepared under the dry atmosphere exhibited higher activity than that prepared under ambient air and pure CeO$_2$.

Keywords: CeO$_2$; MgO; dry reforming; heterogeneous catalysis; in situ XRD

Citation: Taira, K.; Murao, R. High Dispersion of CeO$_2$ on CeO$_2$/MgO Prepared under Dry Conditions and Its Improved Redox Properties. *Energies* **2021**, *14*, 7922. https://doi.org/10.3390/en14237922

Academic Editor: Wasim Khan

Received: 7 October 2021
Accepted: 23 November 2021
Published: 25 November 2021

Publisher's Note: MDPI stays neutral with regard to jurisdictional claims in published maps and institutional affiliations.

Copyright: © 2021 by the authors. Licensee MDPI, Basel, Switzerland. This article is an open access article distributed under the terms and conditions of the Creative Commons Attribution (CC BY) license (https://creativecommons.org/licenses/by/4.0/).

1. Introduction

CeO$_2$ has been used as catalysts and additives due to its high oxygen storage capacity and moderate basicity [1]. The inherent nature of CeO$_2$ makes it attractive for chemical reactions involving oxygen removal and addition. The transformation of renewable energy to fuel gases often employs CeO$_2$ in the process. The addition of CeO$_2$ to catalysts suppresses the carbon deposition on the catalysts during the steam-reforming reactions of CH$_4$ and tar to form syngas [2,3]. CeO$_2$ also improves the resistance of Ni catalysts against H$_2$S during methanation reaction [4] and reforming reaction [3,5] by accelerating the removal of sulfur from the surface of active species. Chemical looping reaction and thermochemical reaction also proceed over CeO$_2$ [6–9]. Despite its various applications, it is preferable to reduce the total consumption of CeO$_2$. The supply and price of CeO$_2$ are volatile. Further, the environmental load during the mining and processing of rare earth elements is higher than those of base metals [10,11]. It is desired to use less CeO$_2$ to make the catalysts, as well as the process, sustainable.

Various techniques have been employed to reduce the consumption of CeO$_2$ by controlling its size and shape [12–14]. Fine CeO$_2$ nanoparticles of high surface areas are prepared by surfactant-assisted methods [15,16]. The size of CeO$_2$ particles also influences the oxygen mobility on the CeO$_2$ surface [17,18]. The lattice constant of CeO$_2$ increases as the size of nanoparticles decreased, leading oxygen vacancies to be more stable than on the coarse CeO$_2$ particles [13,19,20]. The oxygen mobility on the CeO$_2$ surface was also influenced by the plane indices of the surface. Among the low-index surfaces, CeO$_2$(110) exhibits the highest reducibility while CeO$_2$(111) is the least reducible [21–23]. The difference was attributed to the surface energy of the surface [24,25]. CeO$_2$(111) is more stable than CeO$_2$(110) and CeO$_2$(100) because it is the non-polar, close-packed surface of CeO$_2$. Meanwhile, polar CeO$_2$(100) and less-densely packed CeO$_2$(110) surfaces underwent reconstruction of their surface, forming facets containing reduced CeO$_{2-x}$ surface [12]. Such reconstruction of the surface leads them to be more reducible than $CeO_2(111)$. Reportedly, nanorod composed of $CeO_2(110)$ surface and cubic CeO_2 nanoparticles composed of $CeO_2(100)$ surface exhibited higher catalytic activity than conventional CeO_2 nanoparticles mostly composed of $CeO_2(111)$ surface due to higher oxygen mobility on $CeO_2(110)$ and $CeO_2(100)$ surfaces at low temperatures [21–23]. However, such self-standing CeO_2 particles sinter to ~20 m^2/g or less, corresponding to 50 nm in diameter, after calcined at high temperatures >800 °C. Both nanorods and cubic CeO_2 nanoparticles transform into conventional CeO_2 nanoparticles after being exposed to such high temperatures. Al_2O_3 supports are widely employed to improve the thermal stability of CeO_2 nanoparticles. CeO_2 nanoparticles on Al_2O_3 remain ~10 nm in diameter after the heat treatment at >800 °C under the oxidizing atmosphere [26,27]. Contrary, CeO_2 and Al_2O_3 react to form $CeAlO_3$ and other oxides under reducing conditions [28–30]. Such reactions cause structural changes in the catalysts and sintering of CeO_2 nanoparticles. High-temperature tolerant CeO_2 nanoparticles are necessary to further reduce the consumption of CeO_2 under reducing conditions.

Some studies employ MgO as a support to disperse CeO_2 nanoparticles [31–34]. The melting point of MgO is ~2850 °C, which is higher than that of Al_2O_3 supports [31]. MgO is also known for its stability under reducing conditions [35]. Further, CeO_2 nanoparticles are stabilized on MgO supports without forming any composite oxides [31–34,36]. Partial oxidation of CH_4 was demonstrated over CeO_2/MgO catalysts, suggesting their stability under the reducing conditions [34,37]. However, the average diameter of CeO_2 increased to >5 nm after the calcination at 800 °C. Techniques to stabilize finer CeO_2 nanoparticles are necessary to further reduce the consumption of CeO_2.

MgO support of large surface area would contribute to stabilized CeO_2 nanoparticles. Meanwhile, the stability of MgO depends on the atmosphere to which MgO is subjected [38–42]; relative surface energies of low-index surfaces of MgO vary depending on the humidity. Polar close-packed MgO(111) surface is stabilized by hydroxylation independent of the humidity [38]. On the contrary, non-polar MgO(100) is terminated by OH only under humid conditions since the OH-termination does not contribute to the stabilization as it is on the polar surface. Under humid conditions, therefore, OH-terminated MgO(111) is more stable than OH-terminated or pristine MgO(100). Meanwhile, pristine MgO(100) is more stable than OH-terminated MgO(111) under dry conditions [38]. Further, the sintering of MgO is accelerated in the presence of water vapor [39–42]. These reports suggest that the morphology of MgO depends on the humidity of the atmosphere under which the MgO is prepared. These studies imply that the humidity can influence the catalytic activity of the catalysts containing MgO. However, no research assessed the effect of humidity on the catalytic activity and morphology of CeO_2/MgO.

This research demonstrates that CeO_2/MgO catalysts of fine CeO_2 nanoparticles by controlling the atmosphere during the preparation. Calcination in dry air realized CeO_2 nanoparticles smaller than 3 nm in diameter even after heating at 800 °C. The prepared CeO_2/MgO catalyst outperformed pure CeO_2 for dry reforming reaction although the mass ratio of CeO_2 in CeO_2/MgO was less than 1/5 of pure CeO_2.

2. Materials and Methods

2.1. Catalyst Preparation

MgO was prepared by the thermal decomposition of Mg $(OH)_2$ (FUJIFILM Wako Pure Chemical Corporation, Osaka, Japan, 0.07 µm, >99.9%). Although the thermal decomposition of $Mg(OH)_2$ is a common technique to prepare MgO, this study employed ambient air conditions and dry air conditions to see the effect of humidity. The $Mg(OH)_2$ was dried at 110 °C for 1 h and then calcined at 800 °C for 5 h in ambient air or under a dry gas flow of 20% O_2 and N_2 balance. The samples were heated at a heating rate of 4 °C/min. The flow rate of the dry gas was fixed at 100 cm^3/min for 0.3 g of the $Mg(OH)_2$. The O_2 and N_2 were supplied from gas cylinders (Taiyo Nippon Sanso Corporation, Tokyo, Japan, O_2 > 99.99995%, N_2 > 99.99995%). MgO supports prepared in ambient air and under the dry gas flow are denoted by MgO(air) and MgO(dry), respectively. CeO_2 was deposited on the MgO supports by impregnation. An acetone solution of $Ce(NO_3)_3$ $6H_2O$ (Kanto Kagaku, >98.5%) was added dropwise to the MgO supports. The samples were dried at 60 °C with continuous stirring in ambient air. Then, the samples of MgO(air) and MgO(dry) were calcined at 800 °C for 5 h in the ambient air and under the dry gas flow, respectively. The molar ratio of CeO_2 to MgO was adjusted to 0.01/0.99 or 0.05/0.95 by changing the amount of $Ce(NO_3)_3$ $6H_2O$ used for impregnation. The obtained CeO_2/MgO catalysts were denoted by 0.01- or 0.05-CeO_2/MgO(air) and 0.01- or 0.05-CeO_2/MgO(dry) corresponding to the content of CeO_2 and the atmosphere for the heat treatments. A portion of the MgO was subjected to characterizations without the deposition of CeO_2. A pure CeO_2 catalyst was also prepared via a citrate method as a reference [43]. Aquatic solutions of the $Ce(NO_3)_3$ $6H_2O$ and citric acid monohydrate (Kanto Kagaku, Tokyo, Japan, >99.5%) were prepared separately and mixed into a solution. The molar ratio between $Ce(NO_3)_3$ $6H_2O$ and citric acid monohydrate was kept at $\frac{1}{2}$. The pH of the solution was adjusted to ~7.0 by adding an aquatic solution of ammonia (Kanto Kagaku, Tokyo, Japan, 28.0–30.0% as NH_3) dropwise. After stirring continuously for 1 h at room temperature, the solution was dried on a rotary evaporator at 60 °C to become a viscous gel. The gel was further dried at 120 °C for 3 h and calcined at 800 °C for 5 h in ambient air. The high purity of the obtained CeO_2 was confirmed by X-ray fluorescence (Table S1 can be find in the supplementary materials). Before the reaction, the obtained powdery catalysts were pelletized by a compression molding machine and then crushed to a particle size between 500 and 750 µm by stainless steel sieves.

2.2. Catalyst Characterization

The chemical composition of the catalysts was determined by an inductively coupled plasma optical emission spectrometer (ICP-OES). 0.05 g of the sample was dissolved in a mixture of nitric acid and hydrogen peroxide. The solid residue was filtrated and melted in sodium peroxide or sodium borate. The melt was heated in hydrochloric acid to extract the elements in the sample. The obtained liquids were mixed thoroughly and adjusted to a specific volume in a volumetric flask. The prepared liquid was subjected to ICP-OES (Shimadzu, Kyoto, Japan, ICPS-8100). Specific X-ray of λ = 413.765 nm was used for Ce, and those of λ = 280.270 and 383.231 nm were used for Mg. The ICP-OES detector was calibrated using reference liquids of the elements before the measurements. The composition of each catalyst was calculated from ICP-OES results assuming both Ce and Mg exist as oxides, CeO_2 and MgO, in the catalyst.

The surface areas of the catalysts were determined using the Brunauer–Emmett–Teller (BET) method from the N_2 adsorption isotherm at the temperature of liquid nitrogen (MicrotracBEL, Osaka, Japan, BELSORP MINI X). TEM images of the catalysts were taken as bright-field images with a transmission electron microscope (Thermo Fisher Scientific, MA, USA, Tecnai G2) with an accelerating voltage of 200 kV. Each sample was mounted on a carbon grid and measured on a single-tilt holder. The average particle sizes of the CeO_2 were determined using the procedure described in the supporting information of our previous paper [44]. The projected areas of more than 100 nanoparticles on the TEM images were determined with image analysis software. Then, the sphere equivalent diameter was calculated for each particle. The arithmetic mean was calculated using an obtained frequency distribution of the diameter. The total surface area of the CeO_2 particles was also estimated for each catalyst based on the frequency distribution. The morphology of the catalysts was also observed with a scanning transmission electron microscope (STEM). STEM images of the catalysts were obtained as bright-field images with a scanning electron microscope (Hitachi, Tokyo, Japan, SU9000) with an accelerating voltage of 30 kV. Energy-dispersive X-ray spectroscopy (EDS) mapping images were taken along with STEM images with an EDS detector (Oxford Instruments, Tokyo, Japan, X-max 100LE).

The redox properties of the catalysts were assessed by temperature-programmed reduction (TPR) and temperature-programmed oxidation (TPO) studies. TPR of the samples was carried out under a flow of 5% H_2/Ar. All measurements were conducted using an automated catalyst analyzer (MicrotracBEL, Osaka, Japan, BELCAT II) equipped with a thermal conductivity detector (TCD). For the TPR measurements, 0.05 g of the sample were placed on the bottom of a quartz tube and heated to 500 °C under an Ar flow to remove impurities. The sample was then quenched to 30 °C under a flow of Ar. TPR was conducted from 30 °C to 900 °C under a constant 30 cm^3/min flow of 5% H_2/Ar gas at a heating rate of 5 °C/min. The TPO measurements were conducted using a heating chamber (PIKE Technologies, WI, USA, DiffusIR) equipped with a quadrupole mass spectrometer (QMS) (Pfeiffer Vacuum, Aßlar, Germanny, OmniStar). CO_2 is used as an oxidant gas. For each measurement, 0.0275 g of the sample were heated in a heating chamber to 550 °C under a flow of Ar and reduced for 20 min under a constant 50 cm^3/min flow of 5% H_2/Ar gas. The sample was then cooled to 100 °C under a flow of Ar. TPO was conducted at a heating rate of 20 °C/min from 100 to 740 °C under a constant 50 cm^3/min flow of 5% CO_2/Ar gas. The outlet gas was monitored by the QMS. The QMS signals corresponding to the m/z values of 2 (H_2), 28 (CO), 40 (Ar), and 44 (CO_2) were monitored throughout the reaction. The measurement cycle was ~1.0 s at a dwelling time of 0.2 s for each m/z. The gas composition was determined using the relative intensities of the signals to that of m/z 40 (Ar). The total gas concentration was normalized to 100% for each cycle. The QMS system was calibrated before the measurements. Standard gases were used to calibrate the m/z values of 2 (H_2), 28 (CO), 40 (Ar), and 44 (CO_2). The CO concentration was calculated by subtracting the fragmentation of CO_2 and background. The fragmentation ratio of CO_2 to m/z of 44 and 28 was determined using a flow of CO_2/Ar.

The crystallographic structure of the catalysts was determined using in situ X-ray diffraction (XRD) measurements with an XRD instrument (Rigaku, Tokyo, Japn, RINT-TTR III) equipped with a Co X-ray source and an Fe filter. The scanning range was set from $2\theta = 20$ to 80° with a 0.02° step angle at a scanning rate of 40°/min. The X-ray tube voltage and the current were 45 kV and 200 mA, respectively. The sample catalysts were crushed into powder and mounted on an infrared heating attachment (Rigaku, Tokyo, Japan, Reactor X). The infrared heating attachment was purged by a 5% H_2/N_2 flow of 150 cm^3/min under the pressure of 0.1 Mpa. H_2 was supplied from an H_2 generator (Parker Hannifin, OH, USA, H2PEM-260) and diluted with N_2 supplied from a gas cylinder (Tokyo Koatsu, Tokyo, Japan, >99.99995%). The sample was heated up to 900 °C at a heating rate of 5 °C/min under continuous gas flow. The scanning was performed every 10 °C from 200 to 900 °C. XRD measurements were also performed under ambient air with an XRD instrument (Rigaku, Tokyo, Japan, RINT-TTR III) equipped with a Cu X-ray source and an Ni filter. The scanning range was from $2\theta = 20$ to 70° with a 0.02° step angle at a scanning rate of 1°/min at room temperature.

The oxidation state of Ce was measured on an X-ray photoelectron spectroscopy (XPS) analyzer (ULVAC-Phi, Kanagawa, Japan, Quantum-2000). The XPS analyzer was equipped with a monochromated Al X-ray source and a charge neutralizer. The pass energy and the recording step were controlled to 29.35 eV and 0.125 eV, respectively. The binding energy of an isolated u‴ peak of Ce3d$_{3/2}$ was adjusted to 916.70 eV to correct the peak shift derived from the charge-up of the catalysts [45,46]. C1s peak of adventitious carbon was not used for the correction due to the weak intensity of the peak and the overlap with the peaks of carbonates and Ce4s [47,48].

2.3. Dry Reforming Reaction

The obtained catalysts of particle size 500 to 750 μm were subjected to the dry reforming reaction. All the test reactions were conducted at ambient pressure for 6 h at 800 °C using a tubular flow reactor The composition of the gas was fixed to 25% CH_4, 25% CO_2, and Ar as balance at a flow rate of 100 cm^3/min. The amount of catalyst was 0.1 g for all the reactions, of which space velocity was 60,000 cm^3 hr^{-1} g-cat^{-1}. The inlet and outlet gas compositions were analyzed by a gas chromatograph (Shimadzu, Kyoto, Japan, GC-2014), employing Ar as the carrier gas. The absence of N_2 eliminated peak overlap between CO and the balance gas, which allowed the concentration of the product gases to be estimated correctly even at low CH_4 conversions.

The CH_4 conversions were calculated using a procedure described in our previous report [49], assuming a two-step reaction shown in Equations (1) and (2). We used the reaction rates for Equations (1) and (2) as parameters to reproduce the outlet gas composition by the least-squares method. The calculated reaction rates for Equation (1) were used as the CH_4 conversion rates. All the calculations were performed ignoring the carbon deposition on the catalysts and the C_2 species in the outlet gas (<50 ppm). Therefore, the total flow of CH_4 and CO_2 in the inlet gas was equal to that of CO, CO_2, and CH_4 in the outlet gas during the calculation.

$$CH_4 + CO_2 \rightarrow 2CO + 2H_2 \quad (1)$$

$$H_2 + CO_2 \rightarrow CO + H_2O \quad (2)$$

3. Results and Discussion

3.1. Characterization of the Catalysts

The content of CeO_2 in the catalysts was calculated based on the results of ICP-OES. The obtained values were compared to the nominal values calculated from the amount of nitrates used. The nominal values and the experimental values are almost identical to each other (Table 1). This result suggests that all the catalysts were prepared at the nominal chemical composition as intended.

Table 1. Morphological properties of the prepared catalysts.

	Nominal CeO_2 Content (wt%)	CeO_2 Content (ICP-OES) * (wt%)	BET Area (m^2/g)	CeO_2 Diameter (nm) **	CeO_2 Area (m_2/g) ***
MgO(air)	0	0	33.1	-	-
0.01-CeO_2/MgO(air)	4.1	4.20	30.4	5.3 ± 1.3	3.8
0.05-CeO_2/MgO(air)	18.3	18.31	27.9	6.9 ± 2.0	12.5
MgO(dry)	0	0	97.7	-	-
0.01-CeO_2/MgO(dry)	4.1	4.15	58.9	2.6 ± 1.1	6.2
0.05-CeO_2/MgO(dry)	18.3	18.33	38.5	4.5 ± 1.9	15.8
CeO_2	100.0	100.0	13.5	61.6	13.5

* Contents of CeO_2 and MgO in the catalysts were experimentally determined by ICP-OES. Sum of CeO_2 and MgO was normalized to 100%. ** Arithmetic averages were calculated from TEM measurement results. Standard deviations were described after "±" for each average diameter. The diameter of pure CeO_2 was estimated as an area-weighted average assuming the particles were spheres. *** Calculated assuming that CeO_2 particles are hemisphere attaching their flat planes on MgO.

The crystallographic structure of the prepared catalysts was assessed by XRD. No peaks other than CeO_2 and MgO were observed across all of the XRD patterns (Figure 1). The diffraction patterns of CeO_2 were weaker for CeO_2/MgO catalysts than for that of pure CeO_2 because the content of CeO_2 in CeO_2/MgO was less than 20% in the mass ratio (Table 1). Further, no clear peak shift was observed for peaks of CeO_2 and MgO, suggesting that the formation of the solid solution is small. These results are in good accordance with previous studies [31,34]. The amount of CeO_2 and MgO dissolving with each other was negligible even at 1500 °C. On the other hand, a clear difference was observed between the samples in the peak shape. The broader peaks were observed for the samples prepared in the dry atmosphere than those prepared in the ambient air, suggesting the finer MgO and CeO_2 particles of the samples prepared in the dry atmosphere.

The size of the CeO_2 nanoparticles was also compared by TEM measurement (Figure 2). Small particles stuck on the large particles were assigned as CeO_2 particles based on EDS-mapping results (Figure S1). Figure 3 shows the frequency distribution of the CeO_2 diameter. The results are summarized in Table 1. Small CeO_2 clusters (<3 nm), as well as larger nanoparticles (>3 nm), were observed in all the CeO_2/MgO catalysts as reported by Tinoco et al. [32]. The average diameter of CeO_2 increased as the content of CeO_2 in the catalysts increased; 0.05-CeO$_2$/MgO(air) and 0.05-CeO$_2$/MgO(dry) contained larger CeO$_2$ nanoparticles than 0.01-CeO$_2$/MgO(air) and 0.01-CeO$_2$/MgO(dry), respectively (Table 1). Meanwhile, smaller CeO$_2$ nanoparticles were observed on the catalysts prepared under the dry atmosphere. The 0.01-CeO$_2$/MgO(dry) sample contained 2.6-nm CeO$_2$ on average, which is smaller than 5.3-nm CeO$_2$ for 0.01-CeO$_2$/MgO(air) (Table 1). The results of TEM were in good accordance with the results of XRD; the volume-weighted average of CeO$_2$ nanoparticles was comparable for both TEM and XRD results (Table S2).

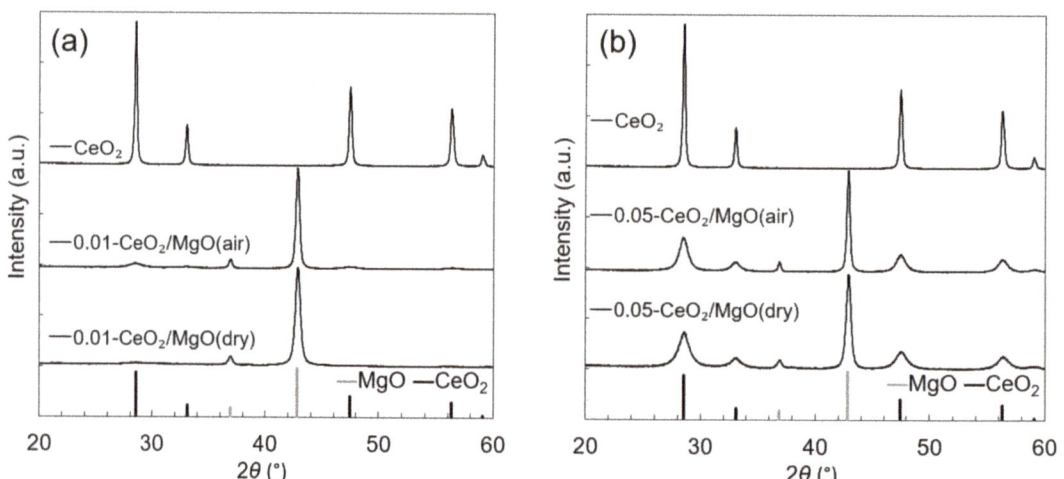

Figure 1. XRD patterns of (**a**) 0.01-CeO$_2$/MgO and (**b**) 0.05-CeO$_2$/MgO catalysts compared to that of pure CeO$_2$. A Cu X-ray source and an Ni filter were used. MgO and CeO$_2$ patterns were assigned based on the references [56,57].

The difference in the size of the CeO$_2$ nanoparticles was attributed to the difference in the BET area. As shown in Table 1, the BET area of MgO(dry) was approximately three times larger than that of MgO(air). MgO(dry) contained small cubic grains caused by the thermal decomposition of Mg(OH)$_2$ (Figure S2) [42]. In contrast, larger octahedral grains were observed in MgO(air) instead of in cubic grains. This difference is ascribed to the water vapor contained in the ambient air. The water vapor accelerates the sintering of MgO of <5 nm in diameter [41,42,50]. Further, the stability of each facet of MgO depends on the humidity of the atmosphere; the {100} face is favored under the dry atmosphere, but the {111} face is under the humid atmosphere [38,51]. The morphological change of MgO due to the humidity caused the sintering of CeO$_2$ nanoparticles during the calcination of the samples at 800 °C.

The precipitous drop in the BET surface areas of 0.01- and 0.05-CeO$_2$/MgO(dry) compared to MgO(dry) was attributed to the condensed nitrate solutions formed during the drying process of the impregnation. During drying, the surface of MgO is covered by Mg(OH)$_2$ since the acetone solution of Ce(NO$_3$)$_3$ contains water; Ce(NO$_3$)$_3$ 6H$_2$O was used as a precursor. Further, the acetone solution contains a small amount of water as an impurity [52]. Therefore, the equilibrium of Equation (3) is present during the drying process.

$$2Ce(NO_3)_3 + 3Mg(OH)_2 \rightleftharpoons 2Ce(OH)_3 + 3Mg(NO_3)_2 \qquad (3)$$

Mg(NO$_3$)$_2$ is soluble both in water and acetone [53,54]. Once the solution of Mg(NO$_3$)$_2$ formed, heterogeneous nucleation of Mg(NO$_3$)$_2$ and Ostwald ripening of MgO would proceed as reported for other oxides [55]. Therefore, the condensed nitrate solution would solve the MgO surface and increase its grain size, which reduced the surface area of MgO.

This result suggests that it is possible to increase the surface area by further optimizing the combination of Ce precursors and solvents.

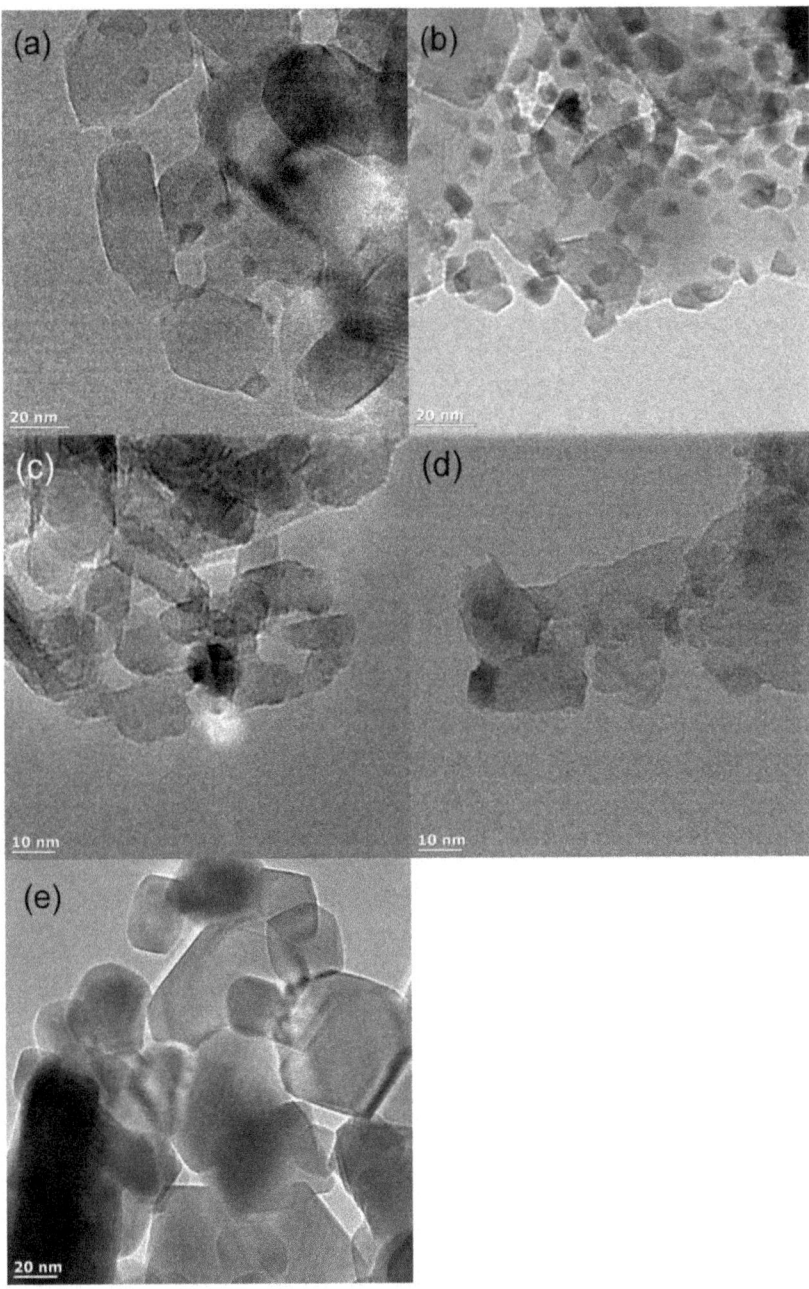

Figure 2. TEM images of (**a**) 0.01-CeO$_2$/MgO(air), (**b**) 0.05-CeO$_2$/MgO(air), (**c**) 0.01-CeO$_2$/MgO(dry), (**d**) 0.05-CeO$_2$/MgO(dry), and (**e**) CeO$_2$. STEM-EDS of the samples are shown in Figure S1.

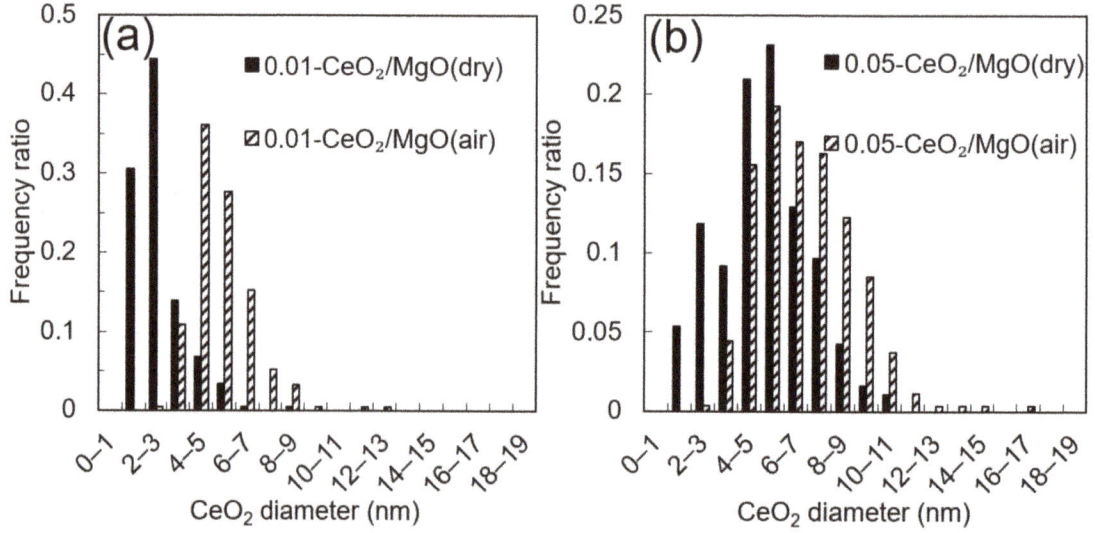

Figure 3. Frequency distribution of the CeO$_2$ diameter in (**a**) 0.01-CeO$_2$/MgO, and (**b**) 0.05-CeO$_2$/MgO.

3.2. Mobility of Oxygen

The mobility of oxygen species on the catalysts was assessed by the TPR measurement (Figure 4). MgO had no peaks because it is irreducible under the measurement conditions (Figure S3). All the other catalysts showed peaks at ~500 and ~800 °C, but the relative intensity of these peaks varied depending on the catalyst. The former and latter peaks are ascribed to the surface capping oxygen and the bulk lattice oxygen of CeO$_2$, respectively [1]. Notably, the intensity of the peaks at ~800 °C lowered as the CeO$_2$ content decreased while the intensity of the peaks at ~500 °C remained comparable, suggesting that the low content of CeO$_2$ led to finer CeO$_2$ nanoparticles (Figure 4a,b). This result is in good agreement with the results of the TEM measurement (Table 1, Figures 2 and 3). Further, the intensity of the peaks at ~800 °C was lower for CeO$_2$/MgO(dry) than for CeO$_2$/MgO(air) for both contents of CeO$_2$. These results suggest that the surface capping oxygen was predominant for the finer CeO$_2$ nanoparticles prepared under the dry atmosphere due to its larger surface area of CeO$_2$ (Table 1). In particular, 0.01-CeO$_2$/MgO(dry) had no peaks at ~800 °C, suggesting that both the surface capping oxygen and the lattice oxygen were removed at ~500 °C. This result is attributed to the fine CeO$_2$ of d = 2.6 nm in 0.01-CeO$_2$/MgO(dry) (Table 1). Such small CeO$_2$ nanoparticles have high reducibility since their unit cell expanded [13,20]. The lattice oxygen of such CeO$_2$ nanoparticles would diffuse ~1 nm from the center of the CeO$_2$ nanoparticles to the surface under the reducing atmosphere.

In addition to the intensity of the reduction peak at ~500 °C, the shape of the peak also changed depending on the preparation condition; a small shoulder appeared at ~350 °C for the samples prepared under dry conditions. This low-temperature peak was attributed to the size-dependent orientation of CeO$_2$ nanoparticles on the MgO surface. Reportedly, the crystallographic orientation of CeO$_2$ nanoparticles varies depending on the size of CeO$_2$; CeO$_2$ of 1–3 nm in diameter faces its {111} surface to the {111} surface of MgO, but CeO$_2$ of 10–20 nm in diameter faces its {100} surface to the {111} surface of MgO [32]. These studies suggest that preferential faceting of CeO$_2$ nanoparticles depends on the size of CeO$_2$. Further, the reducibility of the CeO$_2$ surface strongly depends on their faceting [23]; the reduction peak position shifted from ~500 °C for conventional CeO$_2$ to ~300 °C for nanorod CeO$_2$. Therefore, we considered that the reducibility of the surface capping oxygen was modulated by the difference in the faceting of CeO$_2$ nanoparticles caused by the size variation.

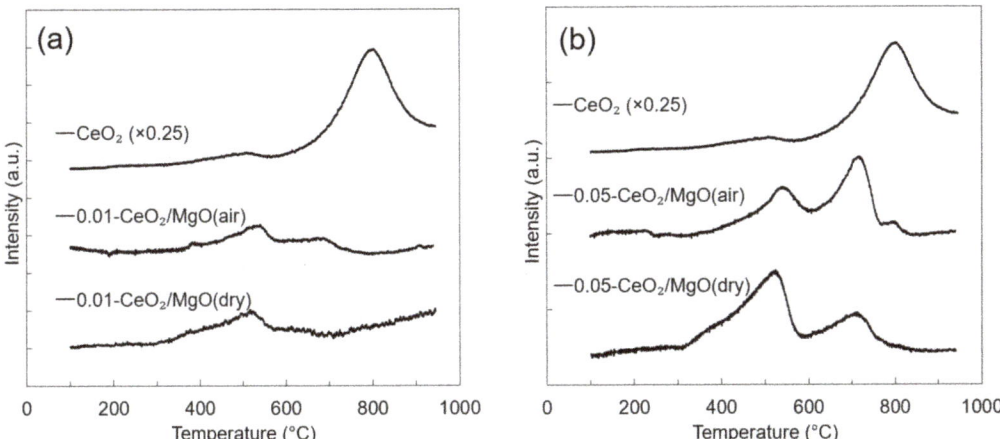

Figure 4. H$_2$-TPR profiles of (a) 0.01-CeO$_2$/MgO and (b) 0.05-CeO$_2$/MgO catalysts compared to that of pure CeO$_2$.

Redox properties of the catalysts were further assessed by TPO using CO$_2$ as an oxidant. The catalysts were reduced in a 5%H$_2$/Ar flow at 550 °C before the TPO measurements. As shown in Figure 5, the formation of CO starts at ~350 °C on all the catalysts. Further, the peak intensity is ~1.5 times higher on 0.05-CeO$_2$/MgO(dry) compared to those on 0.05-CeO$_2$/MgO(air) and pure CeO$_2$. These results are consistent with the H$_2$-TPR (Figure 4b); the reduction peak at ~500 °C on 0.05-CeO$_2$/MgO(dry) is larger than those of the other catalysts. CO$_2$ refilled the oxygen vacancies on the surface that formed during the reduction in an H$_2$ flow. This result also demonstrated the improved oxygen mobility of 0.05-CeO$_2$/MgO(dry) than those of 0.05-CeO$_2$/MgO(air) and pure CeO$_2$. Further, the results shown in Figures 4 and 5 suggest that CeO$_2$/MgO(dry) works as a catalyst for chemical looping combustion (CLC) [6–9]. The CLC proceeds as a two-step reaction; the catalyst is subjected to reducing conditions and then re-oxidized by CO$_2$ or H$_2$O to produce CO or H$_2$, which is identical to the reaction in Figure 5.

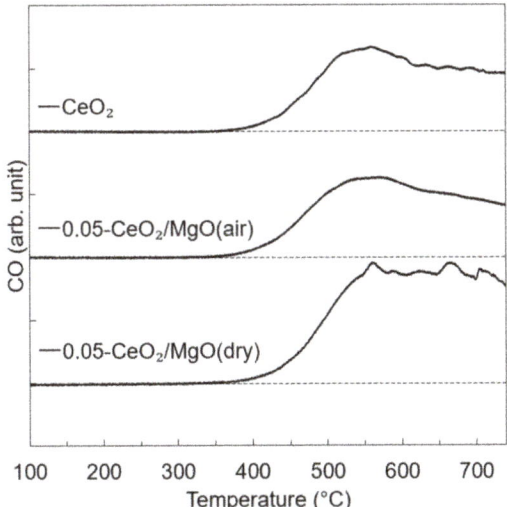

Figure 5. CO$_2$-TPO profiles of the reduced catalysts. The catalysts were heated at 20 °C/min in a 5% CO$_2$/Ar flow after the reduction at 550 °C in a 5% H$_2$/Ar flow.

XPS measurement was also performed on 0.01-CeO_2/MgO(air) and 0.01-CeO_2/MgO(dry) to clarify the difference of CeO_2 nanoparticles caused by the preparation condition. As shown in Figure 6, a distinct difference was observed in the spectra. There were ten peaks derived from Ce^{4+} and Ce^{3+}. The peaks annotated by v, v'', v''', u, u'', and u''' were attributed to Ce^{4+}. Meanwhile, v_0, v', u_0, and u' were due to Ce^{3+}. The binding energies of each peak are described in the caption of Figure 6. Intense peaks of Ce^{3+} denoted by v' and u' were observed for 0.01-CeO_2/MgO(dry). Further, the relative intensity of the $Ce3d_{3/2}$ peak denoted by u''' was lower for 0.01-CeO_2/MgO(dry) than for 0.01-CeO_2/MgO(air), suggesting the presence of Ce^{3+} in 0.01-CeO_2/MgO(dry) [45,46]. Reportedly, the relative intensity of u''' is not linearly correlated to the content of Ce^{3+}, but such a clear difference in the intensity of u''' suggests that more than 30% of Ce in 0.01-CeO_2/MgO(dry) was Ce^{3+} [46,58]. This difference in the Ce^{3+} content is ascribed to the size of CeO_2. Several studies reported that CeO_2 nanoparticles smaller than 3 nm in diameter were relaxed in their crystal structure [13,19]; such small CeO_2 nanoparticles were vulnerable to in situ reductions during the exposure to X-ray under vacuum. As shown in Figure 3, 0.01-CeO_2/MgO(dry) contained a lot of CeO_2 nanoparticles smaller than 3 nm, which would be reduced during the XPS measurement. Such high reducibility of 0.01-CeO_2/MgO(dry) matched well with the results of H_2-TPR (Figure 4). 0.01-CeO_2/MgO(dry) was reduced at a lower temperature than 0.01-CeO_2/MgO(air). Here, it is noteworthy that the effect of precursors on the valence of Ce was minor because both samples were prepared using the same precursor [13].

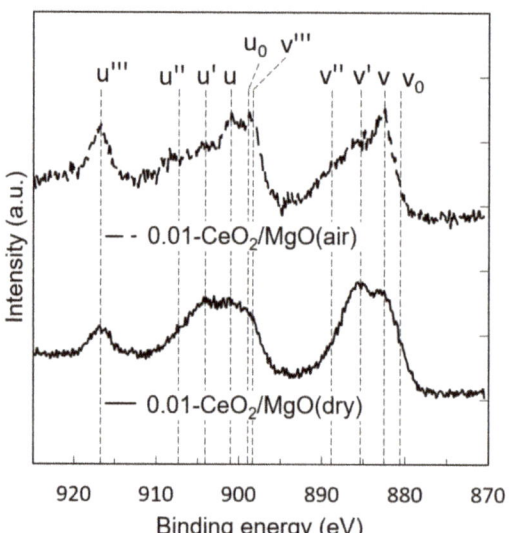

Figure 6. XPS spectra of 0.01-CeO_2/MgO(air) and 0.01-CeO_2/MgO(dry). The peak positions of v_0, v, v', v'', v''', u_0, u, u', u'', and u''' are 880.60, 882.60, 885.45, 888.85, 898.40, 898.90, 901.05, 904.05, 907.45, and 916.70 eV, respectively. The binding energies of all the peaks are identical to the values of the reference [46].

3.3. In Situ XRD Study under the Reducing Atmosphere

In situ XRD was performed to visualize the structural variation of CeO_2 nanoparticles during the reduction. Instead of 0.01-CeO_2/MgO, 0.05-CeO_2/MgO(air) and 0.05-CeO_2/MgO(dry) were used to increase the peak intensity to perform further analyses. The measurements were performed under the same condition as that of H_2-TPR, but an N_2 balance was used as a balance instead of Ar. CeO_2 was also subjected to the same measurement as a reference. The results are shown in Figure 7. Only the peaks of CeO_2 and MgO were observed in the patterns since the phase change from CeO_2 to Ce_2O_3 is slow. The absence of Ce_2O_3 was consistent with the previous studies [1,59,60]. As shown in the patterns of CeO_2/MgO (Figure 7b,c), the peaks of MgO linearly shifted to lower 2θ as the temperature increased, suggesting a continuous thermal expansion of MgO crystals. The peaks of CeO_2 also shifted to the same direction; however, the peaks showed the inflection points at ~650 °C (Figure 7b,c). This implies that structural relaxation was caused by the removal of lattice oxygen of the CeO_2 nanoparticles. Additionally, CeO_2 in both CeO_2/MgO(air) and CeO_2/MgO(dry) underwent structural change at lower temperatures than pure CeO_2 (Figure 7a–c). The inflection point appeared at ~800 °C for pure CeO_2 (Figure 7a). This result suggests that CeO_2 in CeO_2/MgO was reduced faster than pure CeO_2, which is in good agreement with H_2-TPR (Figure 4).

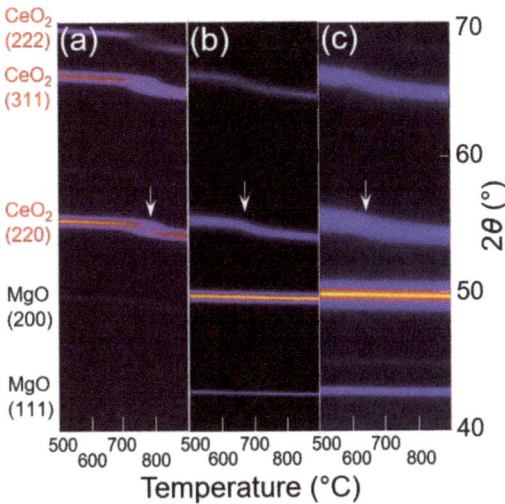

Figure 7. In situ XRD patterns of (a) CeO_2, (b) 0.05-CeO_2/MgO(air), and (c) 0.05-CeO_2/MgO(dry). A Co X-ray source and an Fe filter were used for the measurements. White arrows indicate the inflection points observed on the (220) peaks of CeO_2. The unassigned weak peaks were attributed to the diffraction of Co K-β.

Another difference between the CeO_2/MgO and the pure CeO_2 was observed in the shape of the diffraction patterns of CeO_2 (Figure 7a–c). Notably, the splitting of diffraction patterns was observed for pure CeO_2 around 2θ = ~55° and ~65° at ~750 °C (Figure 6a and Figure S4). This result suggests that the reduction proceeded heterogeneously in CeO_2 grains. As shown in Figure 4, H_2-TPR of pure CeO_2 detected a long tailing to a high temperature at >800 °C. Such tailing means that the reduction of pure CeO_2 was limited by the diffusion of oxygen in CeO_2 [61–63]. Such diffusion-controlled reduction would render the outer side of CeO_2 particles more reduced than the inner part of them. In contrast, such splitting was not observed for CeO_2/MgO catalysts. The reduction of CeO_2 nanoparticles in CeO_2/MgO proceeded more homogeneously than pure CeO_2 due to the small size of CeO_2 nanoparticles.

The variation of unit cell parameter a_0 of CeO_2 was calculated assuming that the structural change of CeO_2 nanoparticles was isotropic. The calculated values of a_0 were normalized by the values at 200 °C to see the variation depending on the temperature. As shown in Figure 8, the unit cell parameter a_0 increased at a lower temperature for 0.05-CeO_2/MgO(dry) than for 0.05-CeO_2/MgO(air). The first derivative of the unit cell parameter held a peak at 640 °C and 680 °C for 0.05-CeO_2/MgO(dry) and 0.05-CeO_2/MgO(air), respectively (Figure S5). This difference was ascribed to the smaller size of CeO_2 in 0.05-CeO_2/MgO(dry). As confirmed by H_2-TPR (Figure 4), 0.05-CeO_2/MgO(dry) was reduced at a lower temperature than 0.05-CeO_2/MgO(air) due to the higher dispersion of CeO$_2$ (Figure 3). The faster removal of the oxygen from the lattice led to the structural change at a lower temperature. In addition, structural relaxation proceeds more easily as the size of CeO$_2$ decreases [13,17,19]. The results of in situ XRD also demonstrated that 0.05-CeO$_2$/MgO(dry) was reduced at lower temperatures than 0.05-CeO$_2$/MgO(air).

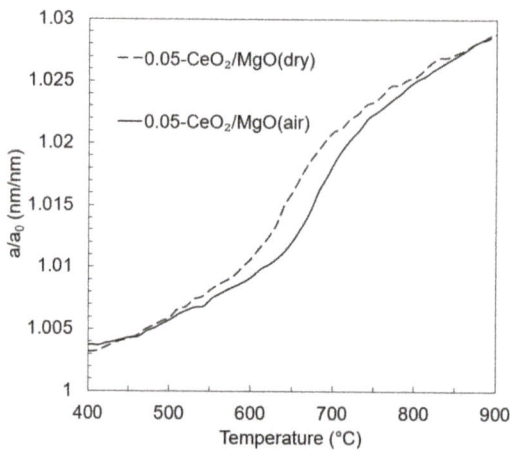

Figure 8. Unit cell parameter a_0 of CeO$_2$ nanoparticles of 0.05-CeO$_2$/MgO(air) and 0.05-CeO$_2$/MgO(dry) calculated from in situ XRD measurement. The vertical axis is normalized by the value of a_0 at 200 °C.

3.4. Dry Reforming Reaction

As shown in previous sections, CeO$_2$ nanoparticles in 0.05-CeO$_2$/MgO(dry) were smaller than those in 0.05-CeO$_2$/MgO(air). Such small CeO$_2$ nanoparticles led to the high mobility of the lattice oxygen. In this section, the difference of the catalysts was demonstrated via a dry reforming reaction, in which the catalysts were exposed to dry reductive gas at high temperatures. Figure 9 shows the average reaction rate of CH$_4$ throughout the reaction. The time course profile of the CH$_4$ conversion rate is also shown in Figure S6. All the catalysts remained white or yellow after the reaction, suggesting the absence of carbon deposition. Both 0.05-CeO$_2$/MgO(air) and 0.05-CeO$_2$/MgO(dry) outperformed pure CeO$_2$ despite their low content of CeO$_2$, i.e., 18.3 wt%. Further, a higher conversion was attained over 0.05-CeO$_2$/MgO(dry) than over 0.05-CeO$_2$/MgO(air). These differences were ascribed to the dispersion of CeO$_2$ nanoparticles. The average diameter of pure CeO$_2$ was 61.6 nm while those of 0.05-CeO$_2$/MgO(air) and 0.05-CeO$_2$/MgO(dry) were 6.9 nm and 4.5 nm, respectively (Table 1). The finer CeO$_2$ nanoparticles led to the higher surface area of CeO$_2$, resulting in the higher catalytic activity. In addition, 0.05-CeO$_2$/MgO(air) exhibited higher activity than pure CeO$_2$ although the area of CeO$_2$ in 0.05-CeO$_2$/MgO(air), 12.5 m^2/g, was smaller than that of pure CeO$_2$, 13.5 m^2/g (Table 1). Our previous study also suggests that MgO promotes CO$_2$ supply to the CeO$_2$ surface due to the strong basicity of the MgO [36]. Such interaction would also contribute to the higher catalytic activity of CeO$_2$/MgO.

The selectivity of the reaction is also illustrated in Figure S7 as the time course profile of the outlet gas composition. All the reaction proceeds under the reaction condition of a low conversion rate of CH$_4$, ~4%. Due to the low conversion rate, the selectivity of H$_2$ against CO was as low as 0.1–0.2 (Figure S7). Notably, the average value of H$_2$/CO throughout the 6-h reaction was slightly smaller for pure CeO$_2$, 0.11, than 0.05-CeO$_2$/MgO(air), 0.18, and 0.05-CeO$_2$/MgO(dry), 0.17. The origin of this difference is uncertain, but we attributed it to the CH$_4$ conversion rate and reducibility of CeO$_2$. As shown in Figure 9, the CH$_4$ conversion rate over the pure CeO$_2$ is lower than those over 0.05-CeO$_2$/MgO. This means the total supply of H$_2$ is less over CeO$_2$ than over 0.05-CeO$_2$/MgO, causing the difference in the H_2/CO ratio. Another possible origin of the difference is the reducibility of CeO_2. H_2-TPR demonstrated that reduction of pure CeO_2 did not complete even at 800 °C (Figure 4). This result suggests that the pure CeO_2 continued to be reduced by the product H_2 during the reaction. Such a reaction would lead to a lower H_2/CO ratio due to the consumption of H_2.

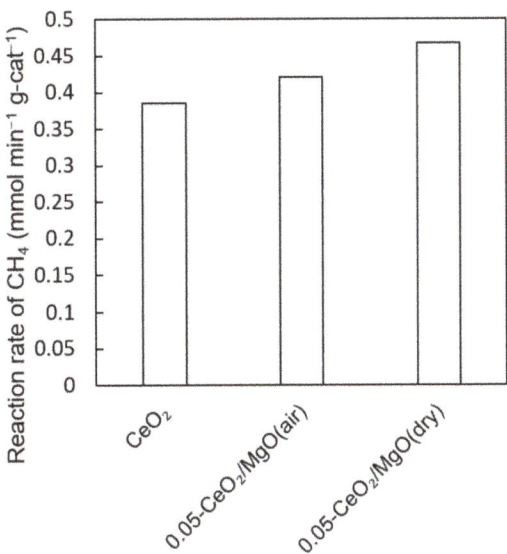

Figure 9. Average reaction rates of CH_4 of 6-h dry reforming reaction over CeO_2 and CeO_2/MgO.

The temporal variation of the CH_4 conversion rate also depended on the catalysts. The pure CeO_2 was deactivated rapidly within the initial two hours but then gradually (Figure S6). On the contrary, such an initial rapid drop of the CH_4 conversion rate was not observed for 0.05-CeO_2/MgO(air) and 0.05-CeO_2/MgO(dry). This difference in the initial behavior was attributed to the stoichiometry of the CeO_2 surface influenced by the diffusion of oxygen in the CeO_2 particles. As shown in Figure 4, the reduction of pure CeO_2 at 800 °C was diffusion controlled. The reduction did not complete even at 900 °C. This result suggests that it takes a long time to reach the equilibrium in the oxygen content of the pure CeO_2 during the reaction. Contrary, the reduction of 0.05-CeO_2/MgO(air) and 0.05-CeO_2/MgO(dry) was almost completed at 800 °C in H_2-TPR (Figure 4), suggesting that it takes a short time to reach the equilibrium in the oxygen content of CeO_2/MgO during the reaction. Therefore, there would be fewer oxygen vacancies on the surface of pure CeO_2 than on CeO_2 nanoparticles in CeO_2/MgO. Such fewer oxygen vacancies of the pure CeO_2 would improve the initial activity because previous studies reported that the C-H activation energy on CeO_2 decreased as the number of oxygen vacancies decreased [64,65].

Notably, 0.05-CeO_2/MgO(air) exhibited steady catalytic activity throughout the 6-h reaction, but 0.05-CeO_2/MgO(dry) was gradually deactivated (Figure S6). The gradual deactivation of 0.05-CeO_2/MgO(dry) was attributed to the gradual sintering of CeO_2 nanoparticles caused by the moisture content in the gas. Dry reforming produces H_2O along with H_2 and CO. Although the concentration of water in the product gas was lower than that of ambient air (Table S3), 0.05-CeO_2/MgO(dry) was exposed to the water vapor at 800 °C during the reaction, causing the structural change of 0.05-CeO_2/MgO(dry). On the contrary, 0.05-CeO_2/MgO(air) was already heated at 800 °C in the ambient air containing ~3 vol% of H_2O before the reaction (Table S3). Therefore, the byproduct H_2O in the product gas affected 0.05-CeO_2/MgO(dry) more than 0.05-CeO_2/MgO(air).

As discussed above, 0.05-CeO$_2$/MgO(dry) gradually deactivated during the dry reforming reaction while 0.05-CeO$_2$/MgO(air) exhibited stable activity. This difference in the behavior of 0.05-CeO$_2$/MgO(air) and 0.05-CeO$_2$/MgO(dry) suggests that the preparation of catalysts under the dry atmosphere is effective only for the reactions in the absence of H$_2$O. Solar-thermochemical reaction is an example of such reactions [7]. The solar thermochemical reaction proceeds as a two-step reaction; the catalyst is subjected to high temperatures to be reduced thermodynamically and then re-oxidized by CO$_2$ at low temperatures to produce CO. The CeO$_2$/MgO(dry) would be suitable for the reaction since it demonstrated high stability and reducibility under dry conditions (Figures 4 and 5).

4. Conclusions

This study demonstrated that stable CeO$_2$ nanoparticles were deposited on MgO by the simple impregnation method. When CeO$_2$/MgO was prepared under the dry atmosphere, the CeO$_2$ nanoparticles remained ~3 nm in diameter even after being heated at 800 °C, which is much smaller than ~5 nm of CeO$_2$/MgO prepared under ambient air. The difference was attributed to the higher surface area of the catalysts prepared under the dry atmosphere. H$_2$-TPR, CO$_2$-TPO, XPS, and in situ XRD showed that CeO$_2$/MgO(dry) exhibited higher oxygen mobility than CeO$_2$/MgO(air) due to the higher dispersion of CeO$_2$ nanoparticles. The higher catalytic activity of CeO$_2$/MgO(dry) than CeO$_2$/MgO(air) was also demonstrated by the dry reforming reaction.

Supplementary Materials: The following are available online at https://www.mdpi.com/article/10.3390/en14237922/s1, Table S1: Chemical composition of CeO$_2$ estimated by X-ray fluorescence, Figure S1: Scanning Transmission Electron Microscope images (a,d) and corresponding energy dispersive X-ray spectroscopy mapping of Ce Lα (b,e) and Mg Kα (c,f) for Ce$_{0.01}$Mg$_{0.99}$O$_{1.01}$ (a–c) Ce$_{0.05}$Mg$_{0.95}$O$_{1.05}$ (d–f), Table S2: Volume-weighted average diameter of CeO$_2$ nanoparticles estimated from TEM and XRD results, Figure S2: TEM images of (above) MgO(air) and (bottom) MgO(dry), Figure S3: H$_2$-TPR profile of MgO(air), Figure S4: XRD patterns of CeO$_2$ during in situ XRD measurement, Figure S5: The first derivative of a0/a0 shown in Figure 8 of the main manuscript, Figure S6: Time course profiles of the CH$_4$ conversion rates during dry reforming reactions, Figure S7. Time course profiles of the outlet gas composition during dry reforming reactions, Table S3: Compositions of the inlet and outlet gases during the dry reforming reaction.

Author Contributions: Conceptualization, K.T.; methodology, K.T.; validation, K.T. and R.M.; formal analysis, K.T. and R.M.; investigation, K.T. and R.M.; resources, K.T.; data curation, K.T.; writing—original draft preparation, K.T.; writing—review and editing, K.T.; visualization, K.T.; supervision, K.T.; project administration, K.T.; funding acquisition, K.T. All authors have read and agreed to the published version of the manuscript.

Funding: This research received no external funding.

Acknowledgments: I would like to show my appreciation to Hiroshima and Tani from Nippon Steel Corporation for the maintenance and troubleshooting of TEM measurements.

Conflicts of Interest: The authors declare no conflict of interest.

References

1. Trovarelli, A. Catalytic properties of ceria and CeO$_2$-Containing materials. *Catal. Rev.-Sci. Eng.* **1996**, *38*, 439–520. [CrossRef]
2. Yang, X.; Da, J.; Yu, H.; Wang, H. Characterization and performance evaluation of Ni-based catalysts with Ce promoter for methane and hydrocarbons steam reforming process. *Fuel* **2016**, *179*, 353–361. [CrossRef]
3. Savuto, E.; Navarro, R.M.; Mota, N.; Di Carlo, A.; Bocci, E.; Carlini, M.; Fierro, J.L.G. Steam reforming of tar model compounds over Ni/Mayenite catalysts: Effect of Ce addition. *Fuel* **2018**, *224*, 676–686. [CrossRef]
4. Gac, W.; Zawadzki, W.; Rotko, M.; Słowik, G.; Greluk, M. CO$_2$ Methanation in the Presence of Ce-Promoted Alumina Supported Nickel Catalysts: H$_2$S Deactivation Studies. *Top. Catal.* **2019**, *62*, 524–534. [CrossRef]
5. Yeo, T.Y.; Ashok, J.; Kawi, S. Recent developments in sulphur-resilient catalytic systems for syngas production. *Renew. Sustain. Energy Rev.* **2019**, *100*, 52–70. [CrossRef]
6. Nair, M.M.; Abanades, S. Tailoring Hybrid Nonstoichiometric Ceria Redox Cycle for Combined Solar Methane Reforming and Thermochemical Conversion of H$_2$O/CO$_2$. *Energy Fuels* **2016**, *30*, 6050–6058. [CrossRef]

7. Arifin, D.; Weimer, A.W. Kinetics and mechanism of solar-thermochemical H_2 and CO production by oxidation of reduced CeO_2. *Sol. Energy* **2018**, *160*, 178–185. [CrossRef]
8. Zheng, Y.; Li, K.; Wang, H.; Tian, D.; Wang, Y.; Zhu, X.; Wei, Y.; Zheng, M.; Luo, Y. Designed oxygen carriers from macroporous $LaFeO_3$ supported CeO_2 for chemical-looping reforming of methane. *Appl. Catal. B Environ.* **2017**, *202*, 51–63. [CrossRef]
9. Galvita, V.V.; Poelman, H.; Bliznuk, V.; Detavernier, C.; Marin, G.B. CeO_2-Modified Fe_2O_3 for CO_2 Utilization via Chemical Looping. *Ind. Eng. Chem. Res.* **2013**, *52*, 8416–8426. [CrossRef]
10. Zaimes, G.G.; Hubler, B.J.; Wang, S.; Khanna, V. Environmental life cycle perspective on rare earth oxide production. *ACS Sustain. Chem. Eng.* **2015**, *3*, 237–244. [CrossRef]
11. Haque, N.; Hughes, A.; Lim, S.; Vernon, C. Rare earth elements: Overview of mining, mineralogy, uses, sustainability and environmental impact. *Resources* **2014**, *3*, 614–635. [CrossRef]
12. Lin, Y.; Wu, Z.; Wen, J.; Poeppelmeier, K.R.; Marks, L.D. Imaging the atomic surface structures of CeO_2 nanoparticles. *Nano Lett.* **2014**, *14*, 191–196. [CrossRef]
13. Paun, C.; Safonova, O.V.; Szlachetko, J.; Abdala, P.M.; Nachtegaal, M.; Sa, J.; Kleymenov, E.; Cervellino, A.; Krumeich, F.; Van Bokhoven, J.A. Polyhedral CeO_2 nanoparticles: Size-dependent geometrical and electronic structure. *J. Phys. Chem. C* **2012**, *116*, 7312–7317. [CrossRef]
14. Matei-Rutkovska, F.; Postole, G.; Rotaru, C.G.; Florea, M.; Pârvulescu, V.I.; Gelin, P. Synthesis of ceria nanopowders by microwave-assisted hydrothermal method for dry reforming of methane. *Int. J. Hydrog. Energy* **2016**, *41*, 2512–2525. [CrossRef]
15. Laosiripojana, N.; Charojrochkul, S.; Kim-lohsoontorn, P.; Assabumrungrat, S. Role and advantages of H_2S in catalytic steam reforming over nanoscale CeO_2-based catalysts. *J. Catal.* **2010**, *276*, 6–15. [CrossRef]
16. Wu, Z.; Zhang, J.; Benfield, R.E.; Ding, Y.; Grandjean, D.; Zhang, Z.; Ju, X. Structure and Chemical Transformation in Cerium Oxide Nanoparticles Coated by Surfactant Cetyltrimethylammonium Bromide (CTAB): An X-ray Absorption Spectroscopic Study. *J. Phys. Chem. B* **2002**, *106*, 4569–4577. [CrossRef]
17. Zhang, F.; Chan, S.W.; Spanier, J.E.; Apak, E.; Jin, Q.; Robinson, R.D.; Herman, I.P. Cerium oxide nanoparticles: Size-selective formation and structure analysis. *Appl. Phys. Lett.* **2002**, *80*, 127–129. [CrossRef]
18. Natile, M.M.; Boccaletti, G.; Glisenti, A. Properties and reactivity of nanostructured CeO_2 powders: Comparison among two synthesis procedures. *Chem. Mater.* **2005**, *17*, 6272–6286. [CrossRef]
19. Spanier, J.E.; Robinson, R.D.; Zhang, F.; Chan, S.; Herman, I.P. Size-dependent properties of CeO_{2-y} nanoparticles as studied by Raman scattering. *Phys. Rev. B* **2001**, *64*, 245407. [CrossRef]
20. Hailstone, R.K.; DiFrancesco, A.G.; Leong, J.G.; Allston, T.D.; Reed, K.J. A study of lattice expansion in CeO_2 Nanoparticles by Transmission Electron Microscopy. *J. Phys. Chem. C* **2009**, *113*, 15155–15159. [CrossRef]
21. Han, W.Q.; Wen, W.; Hanson, J.C.; Teng, X.; Marinkovic, N.; Rodriguez, J.A. One-dimensional ceria as catalyst for the low-temperature water-gas shift reaction. *J. Phys. Chem. C* **2009**, *113*, 21949–21955. [CrossRef]
22. Zhang, J.; Kumagai, H.; Yamamura, K.; Ohara, S.; Takami, S.; Morikawa, A.; Shinjoh, H.; Kaneko, K.; Adschiri, T.; Suda, A. Extra-Low-Temperature Oxygen Storage Capacity of CeO_2 Nanocrystals with Cubic Facets. *Nano Lett.* **2011**, *11*, 361–364. [CrossRef] [PubMed]
23. Aneggi, E.; Wiater, D.; de Leitenburg, C.; Llorca, J.; Trovarelli, A. Shape-Dependent Activity of Ceria in Soot Combustion. *ACS Catal.* **2014**, *4*, 172–181. [CrossRef]
24. Liu, Z.; Sorrell, C.C.; Koshy, P.; Hart, J.N. DFT Study of Methanol Adsorption on Defect-Free CeO_2 Low-Index Surfaces. *ChemPhysChem* **2019**, *20*, 2074–2081. [CrossRef]
25. Branda, M.M.; Ferullo, R.M.; Causá, M.; Illas, F. Relative stabilities of low index and stepped CeO_2 surfaces from hybrid and GGA + U implementations of density functional theory. *J. Phys. Chem. C* **2011**, *115*, 3716–3721. [CrossRef]
26. Damyanova, S.; Bueno, J.M.C. Effect of CeO_2 loading on the surface and catalytic behaviors of CeO_2-Al_2O_3-supported Pt catalysts. *Appl. Catal. A Gen.* **2003**, *253*, 135–150. [CrossRef]
27. Reddy, B.M.; Rao, K.N.; Reddy, G.K.; Khan, A.; Park, S.E. Structural characterization and oxidehydrogenation activity of CeO_2/Al_2O_3 and $V_2O_5/CeO_2Al_2O_3$ catalysts. *J. Phys. Chem. C* **2007**, *111*, 18751–18758. [CrossRef]
28. Shyu, J.Z.; Weber, W.H.; Gandhi, H.S. Surface characterization of alumina-supported ceria. *J. Phys. Chem.* **1988**, *92*, 4964–4970. [CrossRef]
29. Beniya, A.; Isomura, N.; Hirata, H.; Watanabe, Y. Morphology and chemical states of size-selected Pt n clusters on an aluminium oxide film on NiAl(110). *Phys. Chem. Chem. Phys.* **2014**, *16*, 26485–26492. [CrossRef] [PubMed]
30. Ma, Y.; Ma, Y.; Li, J.; Li, Q.; Hu, X.; Ye, Z.; Wu, X.Y.; Buckley, C.E.; Dong, D. CeO_2-promotion of $NiAl_2O_4$ reduction via $CeAlO_3$ formation for efficient methane reforming. *J. Energy Inst.* **2020**, *93*, 991–999. [CrossRef]
31. Nataliya Bochvar, N.; Liberov, Y.; Fabrichnaya, O.; MSIT®. Ce-Mg-O Ternary Phase Diagram Evaluation. Available online: https://materials.springer.com/msi/docs/sm_msi_r_10_012253_02 (accessed on 26 November 2020).
32. Tinoco, M.; Sanchez, J.J.; Yeste, M.P.; Lopez-Haro, M.; Trasobares, S.; Hungria, A.B.; Bayle-guillemaud, P.; Blanco, G.; Pintado, J.M.; Calvino, J.J. Low-Lanthanide-Content CeO_2/MgO Catalysts with Outstandingly Stable Oxygen Storage Capacities: An In-Depth Structural Characterization by Advanced STEM Techniques. *ChemCatChem* **2015**, *7*, 3763–3778. [CrossRef]
33. Pérez Casero, R.; Gómez San Román, R.; Perrière, J.; Laurent, A.; Seiler, W.; Gergaud, P.; Keller, D. Epitaxial growth of CeO_2 on MgO by pulsed laser deposition. *Appl. Surf. Sci.* **1997**, *109–110*, 341–344. [CrossRef]

34. Paunović, V.; Zichittella, G.; Mitchell, S.; Hauert, R.; Pérez-Ramírez, J. Selective Methane Oxybromination over Nanostructured Ceria Catalysts. *ACS Catal.* **2018**, *8*, 291–303. [CrossRef]
35. Plewa, J.; Steindor, J. Kinetics of reduction of magnesium sulfate by carbon oxide. *J. Therm. Anal.* **1987**, *32*, 1809–1820. [CrossRef]
36. Taira, K. Dry reforming reactions of CH4 over CeO2/MgO catalysts at high concentrations of H2S. 2021, J. Cat. Under revision. *J. Cat.* Under revision.. **2021**.
37. Ferreira, V.J.; Tavares, P.; Figueiredo, J.L.; Faria, J.L. Effect of Mg, Ca, and Sr on CeO_2 based catalysts for the oxidative coupling of methane: Investigation on the oxygen species responsible for catalytic performance. *Ind. Eng. Chem. Res.* **2012**, *51*, 10535–10541. [CrossRef]
38. Geysermans, P.; Finocchi, F.; Goniakowski, J.; Hacquart, R.; Jupille, J. Combination of (100), (110) and (111) facets in MgO crystals shapes from dry to wet environment. *Phys. Chem. Chem. Phys.* **2009**, *11*, 2228–2233. [CrossRef] [PubMed]
39. Holt, S.A.; Jones, C.F.; Watson, G.S.; Crossley, A.; Johnston, C. Surface modification of MgO substrates from aqueous exposure: An atomic force microscopy study. *Thin Solid Film.* **1997**, *292*, 96–102. [CrossRef]
40. Eastman, P.F.; Culter, I.B. Effect of Water Vapor on Initial Sintering of Magnesia. *J. Am. Ceram. Soc.* **1966**, *49*, 526–530. [CrossRef]
41. Ito, T.; Fujita, M.; Watanabe, M.; Tokuda, T. The initial sintering of alkaline earth oxides in water vapor and hydrogen gas. *Bull. Cemical Soc. Japan* **1981**, *54*, 2412–2419. [CrossRef]
42. Green, J. Calcination of precipitated $Mg(OH)_2$ to active MgO in the production of refractory and chemical grade MgO. *J. Mater. Sci.* **1983**, *18*, 637–651. [CrossRef]
43. Peng, C.; Zhang, Z. Nitrate–citrate combustion synthesis of $Ce_{1-x}Gd_xO_{2-x/2}$ powder and its characterization. *Ceram. Int.* **2007**, *33*, 1133–1136. [CrossRef]
44. Taira, K.; Nakao, K.; Suzuki, K.; Einaga, H. SO_x tolerant Pt/TiO_2 catalysts for CO oxidation and the effect of TiO_2 supports on catalytic activity. *Environ. Sci. Technol.* **2016**, *50*, 9773–9780. [CrossRef] [PubMed]
45. Burroughs, P.; Hamnett, A.; Orchard, A.F.; Thornton, G. Satellite structure in the X-ray photoelectron spectra of some binary and mixed oxides of lanthanum and cerium. *J. Chem. Soc. Dalt. Trans.* **1976**, 1686. [CrossRef]
46. Romeo, M.; Bak, K.; El Fallah, J.; Le Normand, F.; Hilaire, L. XPS Study of the reduction of cerium dioxide. *Surf. Interface Anal.* **1993**, *20*, 508–512. [CrossRef]
47. National Institute of Standards and Technology. *NIST X-Ray Photoelectron Spectroscopy Database*; NIST Standard Reference Database Number 20; National Institute of Standards and Technology: Gaithersburg, MD, USA, 2000. [CrossRef]
48. Della Mea, G.B.; Matte, L.P.; Thill, A.S.; Lobato, F.O.; Benvenutti, E.V.; Arenas, L.T.; Jürgensen, A.; Hergenröder, R.; Poletto, F.; Bernardi, F. Tuning the oxygen vacancy population of cerium oxide (CeO_{2-x}, 0 <x <0.5) nanoparticles. *Appl. Surf. Sci.* **2017**, *422*, 1102–1112. [CrossRef]
49. Taira, K.; Sugiyama, T.; Einaga, H.; Nakao, K.; Suzuki, K. Promoting effect of 2000 ppm H_2S on the dry reforming reaction of CH_4 over pure CeO_2, and in situ observation of the behavior of sulfur during the reaction. *J. Catal.* **2020**, *389*, 611–622. [CrossRef]
50. Beruto, D.; Botter, R.; Searcy, A.W. H_2O-Catalyzed Sintering of ~2-nm-Cross-Section Particles of MgO. *J. Am. Ceram. Soc.* **1987**, *70*, 155–159. [CrossRef]
51. Cadigan, C.A.; Corpuz, A.R.; Lin, F.; Caskey, C.M.; Finch, K.B.H.; Wang, X.; Richards, R.M. Nanoscale (111) faceted rock-salt metal oxides in catalysis. *Catal. Sci. Technol.* **2013**, *3*, 900–911. [CrossRef]
52. Gottlieb, H.E.; Kotlyar, V.; Nudelman, A. NMR chemical shifts of common laboratory solvents as trace impurities. *J. Org. Chem.* **1997**, *3263*, 7512–7515. [CrossRef] [PubMed]
53. Sangwal, K. Mechanism of dissolution of MgO crystals in acids. *J. Mater. Sci.* **1980**, *15*, 237–246. [CrossRef]
54. Filley, J.; Ibrahim, M.A.; Nimlos, M.R.; Watt, A.S.; Blake, D.M. Magnesium and calcium chelation by a bis-spiropyran. *J. Photochem. Photobiol. A Chem.* **1998**, *117*, 193–198. [CrossRef]
55. Hu, Y.; Lee, B.; Bell, C.; Jun, Y.-S. Environmentally Abundant Anions Influence the Nucleation, Growth, Ostwald Ripening, and Aggregation of Hydrous Fe(III) Oxides. *Langmuir* **2012**, *28*, 7737–7746. [CrossRef]
56. Karen, P.; Kjekshus, A.; Huang, Q.; Karen, V.L. The crystal structure of magnesium dicarbide. *J. Alloys Compd.* **1999**, *282*, 72–75. [CrossRef]
57. McBride, J.R.; Hass, K.C.; Poindexter, B.D.; Weber, W.H. Raman and X-ray studies of $Ce_{1-x}RE_xO_{2-y}$, where RE=La, Pr, Nd, Eu, Gd, and Tb. *J. Appl. Phys.* **1994**, *76*, 2435–2441. [CrossRef]
58. Holgado, J.P.; Alvarez, R.; Munuera, G. Study of CeO_2 XPS spectra by factor analysis: Reduction of CeO_2. *Appl. Surf. Sci.* **2000**, *161*, 301–315. [CrossRef]
59. Galvita, V.V.; Poelman, H.; Rampelberg, G.; De Schutter, B.; Detavernier, C.; Marin, G.B. Structural and kinetic study of the reduction of $CuO-CeO_2/Al_2O_3$ by time-resolved X-ray diffraction. *Catal. Lett.* **2012**, *142*, 959–968. [CrossRef]
60. Perrichon, V.; Laachir, A.; Bergeret, G.; Fréty, R.; Tournayan, L.; Touret, O. Reduction of cerias with different textures by hydrogen and their reoxidation by oxygen. *J. Chem. Soc. Faraday Trans.* **1994**, *90*, 773–781. [CrossRef]
61. Stan, M.; Zhu, Y.T.; Jiang, H.; Butt, D.P. Kinetics of oxygen removal from ceria. *J. Appl. Phys.* **2004**, *95*, 3358–3361. [CrossRef]
62. Al-Madfaa, H.A.; Khader, M.M. Reduction kinetics of ceria surface by hydrogen. *Mater. Chem. Phys.* **2004**, *86*, 180–188. [CrossRef]
63. Goguet, A.; Meunier, F.C.; Tibiletti, D.; Breen, J.P.; Burch, R. Spectrokinetic Investigation of Reverse Water-Gas-Shift Reaction Intermediates over a Pt/CeO_2 Catalyst. *J. Phys. Chem. B* **2004**, *108*, 20240–20246. [CrossRef]

64. Warren, K.J.; Scheffe, J.R. Role of Surface Oxygen Vacancy Concentration on the Dissociation of Methane over Nonstoichiometric Ceria. *J. Phys. Chem. C* **2019**, *123*, 13208–13218. [CrossRef]
65. Kumar, G.; Lau, S.L.J.; Krcha, M.D.; Janik, M.J. Correlation of Methane Activation and Oxide Catalyst Reducibility and Its Implications for Oxidative Coupling. *ACS Catal.* **2016**, *6*, 1812–1821. [CrossRef]

Review

CO$_2$—A Crisis or Novel Functionalization Opportunity?

Daniel Lach *, Jaroslaw Polanski and Maciej Kapkowski

Institute of Chemistry, Faculty of Science and Technology, University of Silesia, Szkolna 9,
40-006 Katowice, Poland; polanski@us.edu.pl (J.P.); maciej.kapkowski@us.edu.pl (M.K.)
* Correspondence: daniel.lach@us.edu.pl; Tel.: +48-32-259-9978

Abstract: The growing emission of carbon dioxide (CO$_2$), combined with its ecotoxicity, is the reason for the intensification of research on the new technology of CO$_2$ management. Currently, it is believed that it is not possible to eliminate whole CO$_2$ emissions. However, a sustainable balance sheet is possible. The solution is technologies that use carbon dioxide as a raw material. Many of these methods are based on CO$_2$ methanation, for example, projects such as Power-to-Gas, production of fuels, or polymers. This article presents the concept of using CO$_2$ as a raw material, the catalytic conversion of carbon dioxide to methane, and consideration on CO$_2$ methanation catalysts and their design.

Keywords: carbon dioxide (CO$_2$); carbon monoxide (CO); CO$_2$ feedstock; methanation; catalyst; catalysis; photocatalysis; Power-to-Gas; catalyst design; heterogenous catalysts database

1. Introduction

Life is based on carbon compounds. The dependence on coal is immanently integrated with human civilization. As the National Oceanic and Atmospheric Administration (NOAA) reports, in 2000 the annual average CO$_2$ concentration in the atmosphere was 369.71 ppm, in 2010 it was 390.10 ppm, and in 2020 it was 414.24 ppm [1]. The growth trend results from the increasing demand for electricity and heat. Additionally, the share of transport in the economy grows, and the current technologies in the power industry and transport are based on fossil fuels [2]. It is not quite clear whether the increase in the CO$_2$ atmospheric concentration of anthropogenic nature is crucial for the greenhouse effect. However, there is no doubt that phenomena related to the overloading of the atmosphere with CO$_2$ result in such an effect. The opinion that it is the anthropogenic CO$_2$ which threatens the fate of our civilization has increasingly often prevailed [3–6]. Therefore, it is very likely that this human dependence on coal leads to a critical excess of carbon dioxide in the atmosphere.

The management of CO$_2$ has become a key issue in the fuels and energy industry. The legislation related to this issue is the subject of European Union regulations, e.g., the European Union Emissions Trading System (EU ETS) [7] and also the Kyoto Protocol, which took effect recently [8,9]. Work related to fuel engineering and new chemistry based on carbon dioxide as the raw material has become a significant challenge. The fact that the carbon dioxide resources in the environment are becoming greater and greater is, beyond dispute, related to CO$_2$ ecotoxicity and its impact on climate change and the natural environment. Hence CO$_2$ is an easily available and cheap chemical raw material [10].

2. Carbon Dioxide Employment

2.1. CO$_2$ Management—Obligation and Opportunity

The first motif of CO$_2$ management results from the regulations, e.g., of the European Union [11]. Because a positive balance of emission is related to high financial penalties, the possibility of reducing emission is attractive in economic terms. Table 1 presents the CO$_2$ emission for selected economies of the European Union countries.

Citation: Lach, D.; Polanski, J.; Kapkowski, M. CO$_2$—A Crisis or Novel Functionalization Opportunity? *Energies* **2022**, *15*, 1617. https://doi.org/10.3390/en15051617

Academic Editor: Wasim Khan

Received: 28 December 2021
Accepted: 19 February 2022
Published: 22 February 2022

Publisher's Note: MDPI stays neutral with regard to jurisdictional claims in published maps and institutional affiliations.

Copyright: © 2022 by the authors. Licensee MDPI, Basel, Switzerland. This article is an open access article distributed under the terms and conditions of the Creative Commons Attribution (CC BY) license (https://creativecommons.org/licenses/by/4.0/).

Table 1. Reduction of greenhouse gas and carbon dioxide emissions in selected countries of the European Union (UE). Own study based on data from [11–15].

Selected EU States	[1] GHG Emission Reduction by 2030, %	[2] GHG Emission in 2005, CO_2 Equivalent, Mt/yr	GHG Emission Limit in 2030, CO_2 Equivalent, Mt/yr	[3] Averaged Value of CO_2 Share in GHG, %	CO_2 Emission Limit in 2030, Mt/yr
Belgium	35	147.174	95.663		69.834
Luxembourg	40	13.166	7.900		5.767
Netherlands	36	225.725	144.464		105.459
Germany	38	993.712	616.101		449.754
Czech Republic	14	148.874	128.032	73	93.463
Poland	7	412.938	384.032		280.343
Slovakia	12	49.748	43.778		31.958
Lithuania	9	23.668	21.538		15.723
Latvia	6	13.081	12.296		8.976

[1] Greenhouse gases (GHG) emission reduction by 2030 as against this emission in 2005 [11]. [2] Greenhouse gases (GHG) emission in 2005 based on [12]. [3] Averaged value of CO_2 share in greenhouse gases (GHG) based on the UNFCCC (2017) [13] and IPCC (2014) [14] data.

As the most recent regulation on greenhouse gases (GHG) emission reduction [15] stipulates, the European Union member states shall reduce the GHG emission in the years 2021–2030, depending on the country, from 0 to 40% below the 2005 level. Carbon dioxide is the dominating component of greenhouse gases. Depending on the source, its share ranges between 65% and 81% [13,14]. In the near future one should expect the economy to be subordinated to the EU requirements and based on so-called smart carbon footprint management [16–21].

The second motif results from the size of the share of individual emission sources. The highest CO_2 emission is now related to the industry, namely power plants, oil and gas processing, cement production, iron and steel metallurgy, or petrochemical industry [10,22,23]. Figure 1 presents the percentage share of individual industry sectors in their total annual carbon dioxide emission.

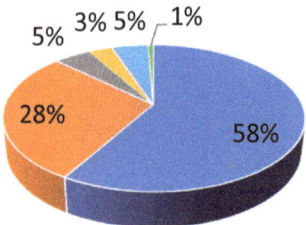

Figure 1. Percentage share of selected industry sources in their total annual CO_2 emissions in the European Union. Data for 2019, extracted from [24].

The anthropogenic impact of carbon dioxide emissions offers a great opportunity for using CO_2 as a raw material or even a feedstock wherever it is currently treated as pollutant or waste. Relatively pure carbon dioxide may be recovered from the production of hydrogen, ammonia, ethylene oxide, gas processing, natural gas liquefaction, hydrocarbons production in the Fischer-Topsch process, or biorefineries, e.g., from ethanol production [20,25]. Such technologies provide a possibility to expand simply the existing plants with units for CO_2 conversion to products useful on the chemical market. Carbon dioxide is now used in the synthesis of urea, salicylic acid, or pigments [10]. In addition, two basic product types may be distinguished, in which CO_2 plays the role of the main raw material. The first type includes inorganic or organic products, the structures of which contain the entire motif of CO_2 molecule. The second type comprises products formed in reactions, in which C-O bonds are broken. This division is of key importance in terms of energy balance and application. The first type of reaction (both inorganic and organic) is not energy-intensive [26] and can frequently proceed spontaneously (at unfavorable kinetics), as in the production of inorganic carbonates. In the case of the second product type, the breaking of C-O bond is energy-intensive and requires the application of reducers, e.g., hydrogen. In the context of smart management of carbon dioxide balance [16–21] it is important that the energy necessary for such reaction would originate from renewable energy sources (solar, wind, geothermal, etc.) or at least from sources different than coal (e.g., nuclear energy). Otherwise, the balance of CO_2 conversion will be reduced by the amount of CO_2 emitted in the process of energy generation, used to carry out the reaction. It is also necessary to remember such factors as the blocking (storage) time of CO_2 molecules in the product. Attention was drawn to this in the report of the Intergovernmental Panel on Climate Change (IPCC) on the capture and storage of carbon dioxide [27]. A long period of use of a product formed from carbon dioxide will block CO_2 for a longer period of time, in this way preventing the reintroduction of carbon dioxide to the atmosphere. In relation to this the first type of product is more stable, e.g., inorganic and organic carbonates, and ensures long-term (from decades to centuries) immobilization of CO_2, while the second type (e.g., fuels or chemicals) immobilizes CO_2 usually for periods of months to a few years. As the second type of product over years may be subject to several cycles of processing and CO_2 releasing (depending on the product life), with the use of renewable energy sources, such technologies are at least equally as attractive as the CCS (carbon capture and storage) technologies [16].

2.2. CO_2 Processing—Examples

The annual conference, "Carbon Dioxide as Feedstock for Fuels, Chemistry and Polymers" (previously known as "CO_2 as Feedstock for Chemistry and Polymers"), in Germany is one of project sources devoted to the employment of CO_2. A few recently proposed strategies for CO_2 use, presented below, originate from there.

The Power-to-Gas (P2G) strategy [28] is a method for carbon dioxide management with good prospects. It consists in using the renewable energy or an energy surplus originating from power plants to produce chemical energy carriers. Figure 2 presents this schematically. Countries in which the power industry is to a large extent based on renewable energy sources (e.g., wind or solar) encounter problems with the energy surplus storage or management [28]. According to the P2G strategy this problem may be resolved by the use of this surplus for water electrolysis, resulting in the origination of hydrogen, which in turn in a reaction with carbon dioxide forms methane or methanol, which are compounds which may be stored and used as an energy source in industry and in the power sector [29,30]. Such processes still require optimizing, increasing the overall productivity, and minimizing the costs. Nevertheless, since 2011 we have been observing a growth of such projects in countries like Germany, Denmark, Switzerland, or Spain [28].

The utilization of CO_2 to produce polymers and chemical compounds is another opportunity for its use [31]. In this way, for example, polyhydroxy alcohols (polyols), polypropylene carbonate (PPC), and cyclic carbonates are obtained. Propylene carbonate and ethylene carbonate is mainly synthesized by catalytic CO_2 cyclization to epoxides [32]. The use of non-toxic and freely available CO_2 not only allows the achievement of compounds with higher added value, but also makes the reaction an example of a green process. Moreover, the reaction is thermodynamically favorable as it uses the high free energy of the epoxides to balance the high thermodynamic stability of the carbon dioxide. However, the differentiation in the rate of the CO_2 cycloaddition reaction depending on the starting substrates, and thus competition with the reaction yielding polycarbonate by-products, requires selective catalysts. Active sites on the catalyst surface are Lewis acids. Therefore, Kelly et al. grafted $ZrCl_4 \cdot (OEt_2)_2$ on the surface of dehydroxylated silica at 700 °C (SiO_2-700) and 200 °C (SiO_2-200) by surface organometal chemistry (SOMC), and tested in the cycloaddition of CO_2 (also from CO_2 from cement factory flue gas) with propylene oxide [33]. As reported by the authors, despite a certain degree of leaching of weakly bound or absorbed zirconium complexes during the first catalysis, the catalyst was active, recoverable, and suitable for reuse in further catalytic cycles. Then, Sodpiban et al. described the heterogeneous catalysts consisting of metal halides ($ZnCl_2$, $SnCl_4$) as active precursors immobilized on the surface silica with ionic liquids that were based on functionalized quaternary ammonium halide salts [34]. The best catalytic systems ($ZnCl_2$(1.99)-IL-I and $SnCl_4$(0.66)-IL-Br) allowed for the practically quantitative conversion of terminal epoxides to the corresponding carbonates under relatively mild conditions (25–40 °C, 1 bar). The catalysts were also tested in a stream of dilute gases of CO_2 (50% CO_2/N_2 mixture) and CO_2 from contaminated sources (20% CH_4 in CO_2 with H_2S as the catalyst poison), obtaining quantitative conversion for the above-described catalysts. The catalysts were deactivated only by the loss of the silica matrix and dehalogenation of quaternary ammonium halide groups with simultaneous poisoning of the active metal centers. Metal-organic frameworks (MOFs) or porous organic polymers (POPs) are new trends in the search for CO_2 cycloaddition catalysts [32]. MOFs are porous crystalline materials with a defined structure and high development of the specific surface area—SSA. On the other hand, POP can be an ideal structure for porphyrin metal (Mg or Al) complexes, giving highly active and selective catalysts under mild conditions [35]. The research carried out in this area allows us to render the financial benefits on market principles. For example, there are already commercial plants producing ethylene carbonate in the reaction of epoxide with carbon dioxide (Asahi Kasei Corporation, Japan). In turn, Novomer, Bayer, or BASF are carrying out investments aimed at implementation of such projects. Breakthrough innovations are expected there [36]. For example, polypropylene carbonate produced with the use of carbon dioxide contains 43 wt% of CO_2. It is biodegradable, stable at high temperatures, flexible, transparent, and features a shape memory effect. This interesting profile of practical properties translates into a wide range of applications. PPC is used in production of packaging foils; foams; softeners; and dispersants for brittle plastics, in particular for originally brittle bioplastics, e.g., polylactic acids (PLA) or polyhydroxyalkanoates (PHA). PPC is frequently used in the production of new materials. PPC combination with PLA or PHA results in obtaining biodegradable, semi-transparent, and easy to process plastics, replacing the widely used acrylonitrile butadiene styrene (ABS). Polyethylene carbonate (PEC) is an equally often studied polymer which employs CO_2. PEC is used as a substitute or additive to traditional plastics made from petroleum. PEC contains 50 wt% of CO_2. Its most interesting practical property consists in the resistance to oxygen transport (permeation), which makes it an interesting packaging material for food. Polyurethane blocks made from polyols, obtained from carbon dioxide, are another example. Such products are used as mattress foams and insulating materials.

Another idea consists in the use of CO_2 as a source of carbon for industrial biotechnologies. In this strategy carbon dioxide is used as food for algae or bacteria [37–40]. In the first case CO_2 feeds cultures of microalgae in special photo-bioreactors or in open ponds. In this case algae may be genetically modified to increase its effectiveness. Biomass is the end product. This method is willingly used to produce various chemicals, in particular in the production of biodiesel and aircraft fuel. The second strategy assumes the use of genetically modified bacteria, which use CO_2 as a source of metabolic carbon, and at the same time as the skeleton to produce specially designed molecules. Modern biotechnology offers already a possibility to "reprogram" bacteria towards synthesis of specified targets. Intensive work continues on modern bacteria strains capable of carbon dioxide consumption and its conversion into specified chemical products [39,40]. An interesting example is the recent research on carbonic anhydrases (CA), enzymes found in algae, archaea, eubacteria, vertebrates, and plants that can convert CO_2 into bicarbonate ions [41]. CA catalyzes the hydration of CO_2, which can finally lead to $CaCO_3$ in the presence of Ca^{2+}. In turn, $CaCO_3$ is already a raw material, e.g., for cement or ceramics. The main advantages of CA include the economically viable sequestration of carbon dioxide and its carbonation at low concentration. However, despite the high catalysis rate, the stability of CA is a significant challenge for its industrial applications. However, these difficulties have been partially overcome by strapping CA on appropriate surfaces, e.g., biochar, alginate, polyurethane foam, or nanostructured materials.

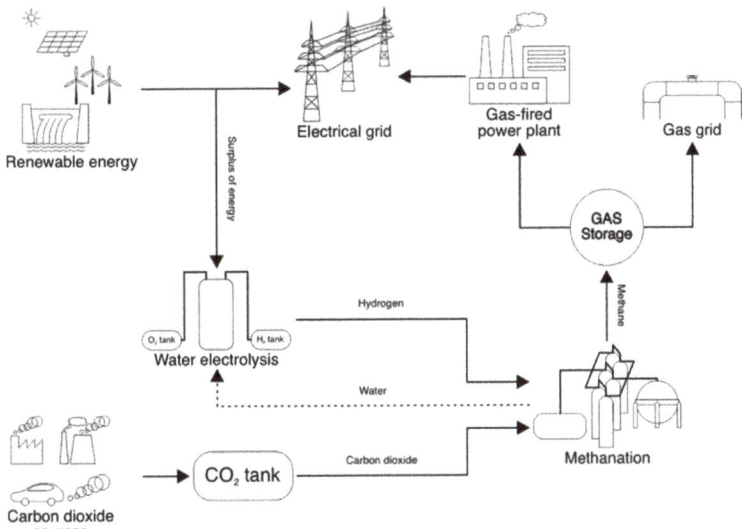

Figure 2. Power-to-Gas strategy. Energy from renewable energy sources is used in electrolysis to produce hydrogen. Hydrogen with carbon dioxide is converted into methane in the CO_2 methanation process. The methane is then stored, released into the gas grid, or used in cogeneration and in gas turbines to produce energy.

Preparing an environmentally friendly solvent and agent with specific properties can also involve carbon dioxide. The subject matter is a supercritical fluid of CO_2 [42]. Such a fluid behaves like gas and liquid at the same time. It is gas-like because it is inviscid and expands to fill a container and liquid-like in terms of density, high heat capacity and conductivity, and solubility. It is non-toxic, non-explosive, thermally stable, and widely available. It is mainly used as a solvent or working fluid. Supercritical CO_2 is an effective solvent for complicated extractions, e.g., nonpolar organic compounds. It does not cause the toxic residual solvent problem, and it is easy to separate/remove from the system. Due to its low critical point, it is an ideal liquid for extracting volatile compounds, compounds with high molecular weight, and compounds with a low degradation temperature. Supercritical CO_2 has proved helpful in the following areas:

- in pharmacy to reduce the particle size of a drug, improving its solubility, and thus bioavailability [43–47],
- in impregnating the compound in pre-formed carrier particles, e.g., of the active compound on the drug carrier [48],
- in micronization and creation of nanoparticles [44,47], and in the development of environmentally friendly dyes [49] and DSSCs (dye-sensitized solar cells) technology [50],
- as an advantage over conventional extraction of, e.g., essential oils from herbs that exhibit various biological, therapeutic, and aromatic properties [51,52].

The critical temperature and pressure of carbon dioxide ($T cr$ = 31.1 °C and pcr = 73.8 bar) are roughly similar to the ambient conditions. Supercritical CO_2 reduces the compression work significantly in the closed-loop compression cycle. Heat dissipation to approximately ambient temperature is observed. Therefore, it is also an attractive working fluid in energy generation technologies and systems, as amply summarized in [53].

However, the use of carbon dioxide would not be possible without an appropriate method of its capture. The equipment of Climeworks company offers an interesting solution [54], which sucks the air containing CO_2 or exhaust gas, and with the involvement of special filters made of porous granulate modified with amines, binds CO_2. After the filter saturation with carbon dioxide it is heated to approx. 100 °C, using low-quality heat as the source of energy. CO_2 is released from the filter and gathered in the form of pure gas, which may be used as a substrate. The air free of carbon dioxide is released into the atmosphere. The cycle is repeated and the applied filters may be used many times, even in a few thousand cycles. This technology may be an important element in the aforementioned concepts, but it is important first of all as an industrial "generator" of clean air. Moreover, the topic of separating CO_2 from gases is being intensively developed even with computational modeling. For example, Ghiasi et al. report that the calculated permeation barrier, selectivity, and thermodynamic functions for CH_4, H_2S, N_2, and CO_2 passing through finite porosity graphene doped with nitrogen atoms indicate a highly efficient and selective material for carbon dioxide separation [55]. In turn, Shaikh et al. describe the reaction mechanism of CO_2 absorption by the amino-acid ionic liquid [56]. They reveal the reaction pathway employing DFT calculations. Using the MD method, they report the cation–anion interaction for two different glycinate-based ionic liquids with structurally similar cations with different alkyl chain lengths. Since the gases for CO_2 recovery are approximately 10% water, the authors also provide simulations with its participation. They note that the interaction between the cation and anion is reduced in the presence of water by reducing the diffusion coefficient of the cation, thus reducing carbon dioxide uptake. Nevertheless, ionic liquid is a promising agent for CO_2 capture, due to the high CO_2 solubility, recycling (almost zero vapor pressure), and fine-tuning dependence on the task.

3. Carbon Dioxide Methanation and Nanocatalysis—The Focal Point in CO_2 Conversion

Catalysis is one of important elements of smart CO_2 management. In particular, many papers have been devoted to catalytic conversion of carbon dioxide to methane. Figure 3 shows an increasing number of papers.

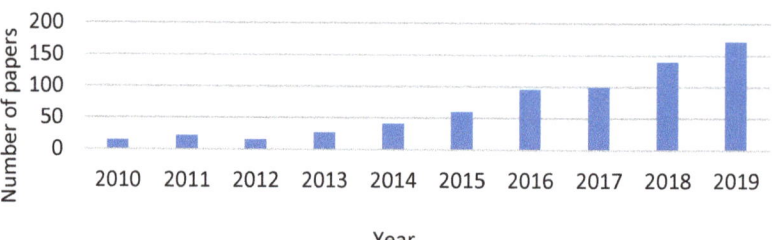

Figure 3. Quantity of publications on catalytic CO_2 methanation from 2010 to 2019. Data from the ISI Web of Science (Thomason Reuters) database. Query conducted for: "catalytic CO_2 methanation".

Catalytic methanation is a central issue of the Power-to-Gas concept [28]. According to statistics, in 2011 the share of papers on CO_2 methanation in all Power-to-X projects (where X is: Gas, Power, Chemicals) was already 27% (Figure 4). The share of catalysis among various CO_2 methanation strategies was 44% and that was the second largest contribution, immediately next to biological methods.

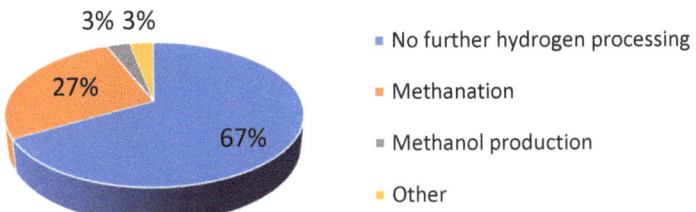

Figure 4. Share of further processing of hydrogen in Power-to-X (X: Gas, Power, Chemicals, Fuels). Data extracted from [28].

3.1. CO₂ Methanation

Carbon dioxide hydrogenation was considered for the first time by Paul Sabatier and Jean B. Senderens in 1902. In the paper "Nouvelles syntheses du methane" [57] they proved that one mole of methane may be obtained in the reaction of one mole of carbon dioxide with four moles of hydrogen, acc. to reaction:

$$CO_2 + 4H_2 \rightleftharpoons CH_4 + 2H_2O$$

This reaction is exothermic and spontaneous. At room temperature (~25 °C) its enthalpy (ΔH) is −165 kJ/mol and the Gibbs free energy (ΔG) is −113.5 kJ/mol [10]. ΔG describes the maximum free energy (energy that can be turned into work) that can be released or adsorbed when it goes from the initial state to the final state. In the CO_2 methanation, a negative ΔG indicates that the substrates (initial state) have more free energy than the products (final state). Therefore, the move towards products involves the release of energy. Energy does not have to be provided for the reaction to occur—it occurs spontaneously. In turn, ΔH refers to the difference between the bond energy of products and substrates. A negative ΔH means a heat release during the reaction towards the products. In the temperature range from 25 to 500 °C, ΔG and ΔH is presented in Figure 5. If the reaction is exo-energetic in one direction, it is also endo-energetic in the opposite direction. Therefore, if the Gibbs free energy in methanation increases rapidly with the rise of temperature (provision of thermal energy), so that above 500 °C it becomes positive, then in the high temperature range, the reverse reaction—methane reforming ($CH_4 + H_2O \rightleftharpoons CO + 3H_2$)—prevails and disturbs the obtaining of methane [58]. However, the course of CO_2 methanation is more complicated and may comprise many intermediate or side reactions. Jiajian Gao specifies them in Table 2 and gives their equilibrium constant K from 200 to 800 °C in Figure 6 [59].

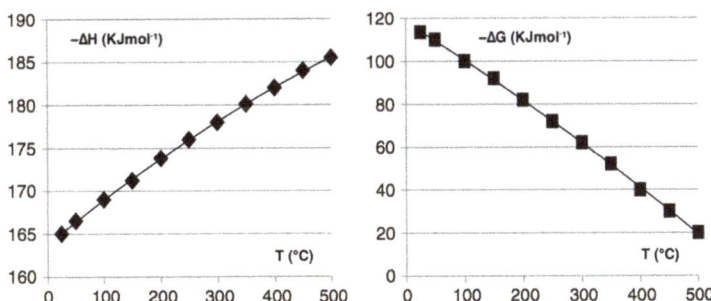

Figure 5. Enthalpy and Gibbs free energy for CO_2 methanation in the temperature range from 25 to 500 °C. Data extracted from [10].

Table 2. Main possible reactions during carbon dioxide methanation. Data extracted from [59].

Reaction Number	Reaction Equation	$\Delta H_{298\,K}$, kJ mol^{-1}	$\Delta G_{298\,K}$, kJ mol^{-1}
R1	$CO + 3H_2 \Longleftrightarrow CH_4 + H_2O$	−206.1	−141.8
R2	$CO_2 + 4H_2 \Longleftrightarrow CH_4 + 2H_2O$	−165.0	−113.2
R3	$2CO + 2H_2 \Longleftrightarrow CH_4 + CO_2$	−247.3	−170.4
R4	$2CO \Longleftrightarrow C + CO_2$	−172.4	−119.7
R5	$CO + H_2O \Longleftrightarrow CO_2 + H_2$	−41.2	−28.6
R6	$2H_2 + C \Longleftrightarrow CH_4$	−74.8	−50.7
R7	$CO + H_2 \Longleftrightarrow C + H_2O$	−131.3	−91.1
R8	$CO_2 + 2H_2 \Longleftrightarrow C + 2H_2O$	−90.1	−62.5

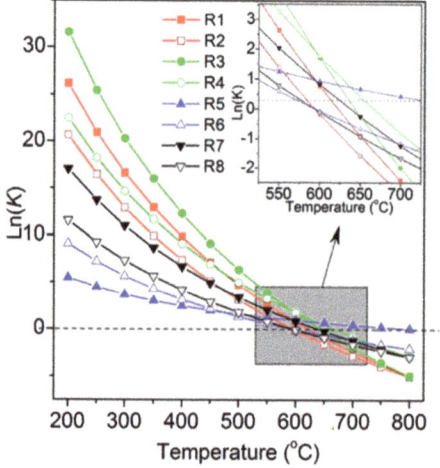

Figure 6. The equilibrium constants (K) for the reactions presented in Table 2, in the temperature range from 200 to 800 °C. © Adopted from [59].

Analysis of the above data can conclude that the temperature is the main parameter affecting the equilibrium. Therefore, from the thermodynamic point of view, the methanation reaction of carbon dioxide should be carried out at low temperatures. However, under such conditions the reaction rate goes down. Hence the CO_2 hydrogenation requires the application of a catalyst [23,60]. It allows the achievement of an acceptable reaction rate and a reduction in the cost of the process itself [61,62].

3.2. Catalyst in Methanation

Metals from group VIII to XI stand out among methanation catalysts [63]. Nickel is probably the most frequently studied metal [64–67]. It features the most favorable ratio of metal price to its activity. Additionally, ruthenium and rhodium show interesting properties [67–71]. In the case of Ru and Rh catalysts, apart from a high activity, their ability to prevent sintering and accumulation of carbon particles is their important advantage, which makes them additionally resistant to deactivation. In addition, Ru stands out in the low-temperature methanation, e.g., in the Ru/TiO$_2$ system [72] or Ru/Ni_nanowires [73]. A low temperature is an important parameter optimizing the thermodynamic and energy efficiency. Numerous studies are related to the possibility of lowering the temperature. Using the example of selective carbon monoxide (CO) methanation [74], Table 3 presents a summary of studies in this field.

Table 3. Profile of selected catalysts in CO methanation. Data extracted from [74].

Catalyst	[1] WHSV, cm^3 g^{-1} h^{-1}	[2] GHSV, h^{-1}	Composition of Inlet Gases, % CO/CO$_2$/H$_2$O/H$_2$	[3] Reaction Characteristic and Yield				
				T_{min}, °C	S_{min}, %	T_{max}, °C	S_{max}, %	mol CO g^{-1} h^{-1}
10% w/w Ni/CeO$_2$	26,000	46,000	1.5/20/10/60	250	89	320	50	0.0160
			1/20/10/60	240	100	315	45	0.0106
			0.5/20/10/60	230	99	290	31	0.0053
	6000	12,000	1/20/10/60	210	100	265	50	0.0025
	13,000	26,000		225	100	280	50	0.0053
	43,000	84,000		265	94	295	54	0.0176
10% w/w Ni/ZrO$_2$	~150,000	-	0.5/14.8/0.8/59.2	280	~90	300	~70	0.0307
1.6% w/w Ni/ZrO$_2$	-	10,000	1.14/21.43/1.8/74.8	260	~60	280	~60	-
10% w/w Ni/TiO$_2$	-	10,000	0.2/16.1/18.4/65.3	200	~80	-	-	-
5% w/w Ru/TiO$_2$	~150,000	-	0.5/14.8/0.8/59.2	220	~70	260	~20	0.0307
5% w/w Ru/TiO$_2$				220	60	260	20	0.0055
5% w/w Ru/ZrO$_2$	27,000	-	0.5/18/15/40	265	80	310	50	0.0055
5% w/w Ru/CeO$_2$				250	75	300	30	0.0055
3% w/w Ru/Al$_2$O$_3$	-	13,500	0.9/24.5/5.7/68.9	220	<50	-	-	-
2% w/w Ru/Al$_2$O$_3$	-	10,000	0.3/4.8/75/18.8	270	<20	-	-	-
30% w/w Ru/CNT				220	-	-	-	0.0059
30% w/w Ru-ZrO$_2$/CNT	12,000	-	1.2/20/0/78.8	180	100	240	35	0.0059
1% w/w Ru/MA-33Ni	-			185	100	245	50	-
1% w/w Ru/MA-40Ni	-	2800	0.9/17/15/67.1	185	100	260	50	-
1% w/w Ru/MA-50Ni	-			195	100	270	50	-

[1] WHSV—weight hourly space velocity (flow of reagents per unit of catalyst mass in the unit of time). [2] GHSV—gas hourly space velocity (volumetric flow of reagents per unit of catalyst volume in the unit of time). [3] T_{min} and T_{max}—minimum and maximum temperature, setting the range in which CO concentration in the reformate is less than 10, or 20 ppm in a few cases. S_{min} and S_{max}—reaction selectivity at T_{min} and T_{max}, respectively.

Another issue is the catalyst activity dependence on the support, on which the selected metal has been placed. For the catalyzed reaction it is favorable to maximize the metal surface area for a specific metal weight [75]. Therefore, small metal particles are synthesized (usually smaller than 1–10 nm), with a narrow size distribution, but with a uniform location on a large specific surface of a thermally stable substrate [23,63,76]. Hence, support in the form of oxides (e.g., SiO$_2$, Al$_2$O$_3$, TiO$_2$), zeolites, carbon, or metaloorganic compounds is distinguished. Support affects also the adsorption and catalytic properties. Figure 7 may be an example, presenting the difference between the selected oxide support of nickel catalyst and the yield of CO$_2$ methanation.

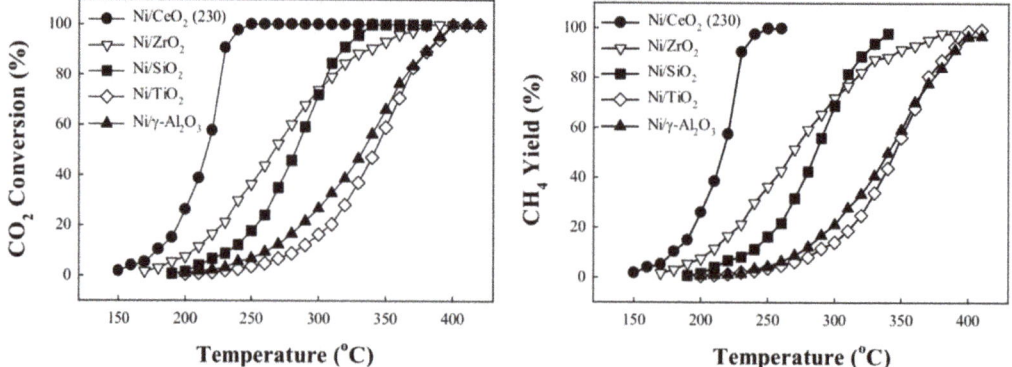

Figure 7. Impact of catalyst supports on the yield of CO_2 to CH_4 conversion. Reaction conditions: 1 mol% CO_2, 50 mol% H_2, 49 mol% He, F/W = 1000 mL/min/g_{cat}. © Adopted from [77].

Studies on the support of methanation catalyst were enhanced with studies on catalytic promoters, that is, substances added to improve or change the catalyst operation. MgO is an example of a catalyst promoter which, introduced to Ni/Al_2O_3 catalyst, increases the thermal stability [78] and resistance to carbon parts precipitation [79]. La_2O_3 increases the Ni/Al_2O_3 catalyst activity via the increase in the nickel dispersion and hydrogen capture [80]. The enhancement of nickel catalyst with V_2O_3 improves its activity, thermal stability, and resistance to sintering [81]. The addition of CeO_2 allows the achievement of a higher susceptibility to reduction and long-term stability [82]. In turn, potassium increases the selectivity towards conversion to higher hydrocarbons [83]. In the context of obtaining methane this is obviously not a desired effect.

The type of support is significant for the CO_2 methanation mechanism [71,84–87]. Hydrogenation of carbon dioxide may proceed via various paths through different structures, which include CO, -OCH_3, and $HCOO^-$ groups. Their origination, further reaction, as well as adsorption and desorption frequently depend on the morphology of the support surface. For example, mezostructural silica, which due to the presence of internal and interparticle pores increases the number of free oxygen sites in the catalyst, is decisive in a particular mechanism of the reaction [88–90]. It is schematically presented in Figure 8. According to this theory, CO_2 and H_2 are adsorbed on the metallic catalyst. As a result of the dissociation of molecular forms, CO, O, and H originate then, which can migrate to the carrier surface. In the next stage CO reacts with oxygen from the carrier surface, forming formate or carbonyl groups in a bridge or bidentate system. In addition, the formation of bidentate formate requires an additional reaction with hydrogen. An oxygen atom is subject to surface stabilization through interaction with electron gaps of the oxide carrier, close to the metal. Oxygen stabilized in this way reacts with hydrogen forming a hydroxyl group, which in a further reaction with hydrogen will form a molecule of water. Oxygen-rich forms of carbon formed on the surface, that is, carbonyl and formate, are hydrogenated to methane.

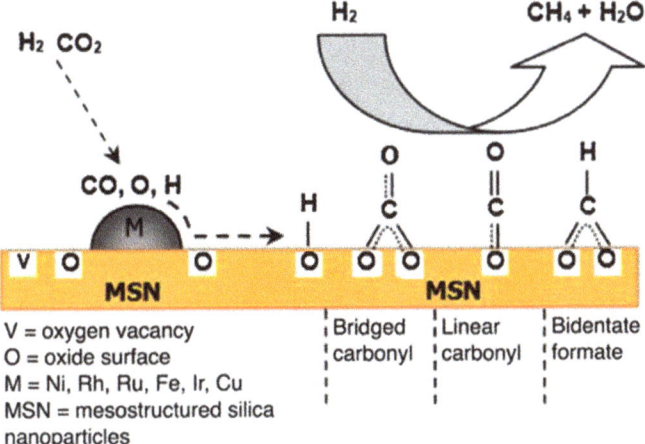

Figure 8. Likely mechanism of CO_2 methanation using the catalyst that is based on mesostructured nanosilica support. © Adopted from [88].

The subsequent essence of matters is the diffusion effect [91,92]. It is a process on the catalyst site that, in a simplified description, may include the following steps: (1) transport of the reactants from the gas phase to the catalyst surface (external diffusion), (2) diffusion of substrates to the surface inside the catalyst pores (internal diffusion), (3) surface operations (chemisorption and catalytic reaction), (4) diffusion of reaction products from inside the catalyst pores to the outside surface (internal diffusion), and (5) migration of reaction products from the catalyst surface to the gas phase (external diffusion). Depending on the morphology of the catalytic surface, the effect of external and internal diffusion is considered. The external diffusion effect depends on the size of the catalyst grains, the flow rate, and the diffusion properties of the reactants. In turn, the internal diffusion effect depends on the porosity of the material, the pore size and distribution, pore connectivity, and the size of the catalytic material grains. The diffusion effect is even more significant when considering the concentration and temperature gradients inside and over the catalyst surface. This topic is discussed in detail in the review [93]. Nevertheless, it is worth noting that this effect is often wrongly ignored, which causes a misinterpretation of the results. Diffusion plays a role in such essential factors as the rate and bottleneck of the reaction or the conversion and product distribution.

The combination of metal and specified support is also frequently studied in the photocatalytic methanation [94]. It was observed that the application of heat and light together can minimize the energy consumption and ensure unique features which cannot be achieved in conventional thermocatalytic reactions [95–97]. Light absorbed by metallic nanoparticles of the catalyst and by reagents existing on their surface is a source of intraband or interband transformations, which generate electrons with a high kinetic energy, so-called hot electrons [97–99]. Hot electrons are effective activators of reagents or intermediate compounds. As a result, a reduced activation energy is observed [100]. For example, in the reaction of carbon dioxide methanation, at 150 °C, hot electrons formed as a result of light absorption by a CO_2 molecule (adsorbed on the metallic surface of Ru/SiO_2 catalyst) increase the conversion of carbon dioxide to methane from 1.6% to 32.6% [101]. Figures 9 and 10 compare Ru/SiO_2 and Rh/SiO_2 catalysts in the CO_2 methanation with the involvement of light and without.

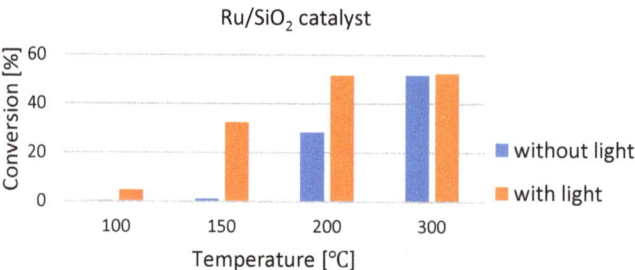

Figure 9. CO_2 conversion on Ru/SiO_2 catalyst with and without light. Data extracted from [101]. Conditions: 0.5% vol. CO_2/N_2 (50 sccm) and H_2 (1.5 sccm). Lamp parameters: Xe 35 mW cm^{-2} with water cooling to exclude the heat effect from the light.

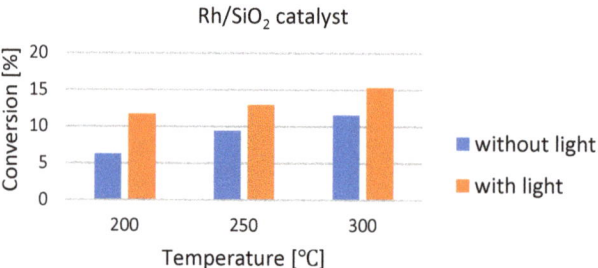

Figure 10. CO_2 conversion on Rh/SiO_2 catalyst with and without light. Data extracted from [101]. Conditions: 0.5% vol. CO_2/N_2 (50 sccm) and H_2 (1.5 sccm). Lamp parameters: Xe 35 mW cm^{-2} with water cooling to exclude the heat effect from the light.

The activity of these catalysts is additionally conditioned by the size of metal nanoparticles (Figure 11). Larger nanoparticles, e.g., ≥5 nm, reduce the activation barrier for CO_2 molecule dissociation on the metal surface. In the case of a photosensitive system this results in a larger number of hot electrons, which improve the reaction kinetics.

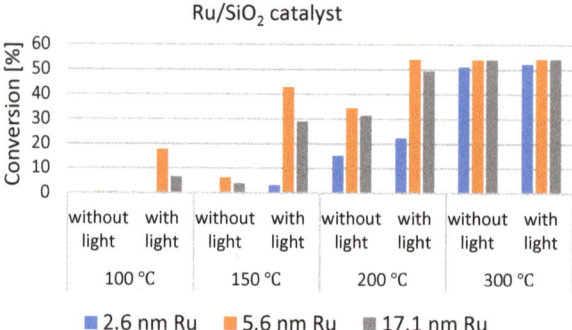

Figure 11. CO_2 conversion on Ru/SiO_2 for different sizes of Ru nanoparticles. Data extracted from [101].

The last issue is the method of catalyst preparation. The selection of preparative method may determine such factors as the size and shape of metal nanoparticles, their uniform distribution on the support, limitation of nanoparticle aggregation, as well as minimization of the used metal [75,102]. Many various methods have been presented in the review entitled "Methods for Preparation of Catalytic Materials" [102]. However, in the context of the aforementioned silica becoming increasingly popular in nanomethods, a proprietary method of our team may draw attention. The method comprises two main stages. The first of them consists in the synthesis of amorphous silica, which plays the role of an intermediate carrier and matrix for metallic nanoparticle generation. The second is the matrix digesting and transferring nanoparticles of the selected metal onto the target support. It is graphically presented in Figure 12. Silica is synthesized by the Stöber method [103]. The aim consists in obtaining spherical, monodisperse, and uniform sizes of silica nanoparticles from the water solution of alcohol and silicon alcoxides at the presence of ammonia as the catalyst. Two basic reactions are distinguished:

$$\text{Hydrolysis: Si-(OR)}_4 + H_2O \rightleftharpoons \text{Si-(OH)}_4 + 4R\text{-OH}$$

$$\text{Condensation: 2Si-(OH)}_4 \rightleftharpoons 2(\text{Si-O-Si}) + 4H_2O$$

Hydrolysis leads to the formation of silanol groups, while siloxane bridges result from the condensation polymerization. The reaction product depends on the type of silicon alcoxide and alcohol. The authors of the methods emphasize that particles prepared in solutions are the smallest, and the particle size increases with the growing length of the alcohol carbon chain. Rao et al. [104] in turn pay attention to the size and deviation of silica grain size through modification of the concentration of silicon alcoxide and alcohol, ammonia concentration, water content, and the change of reaction temperature. This allows the fine-tuning of the physical properties of silica, which is extremely important for later generation of specified sizes of metal nanoparticles. The second stage comprises nanometal growing on the matrix, reducing the intermediate conjugate (metal-silica) with hydrogen, digesting the silica with lye (when other support is needed), transferring metallic nanoparticles onto the surface of the target support, or separating metal nanoparticles. This method allows for nanomanipulation of nanoparticles' size and shape, reduction of their tendency to aggregate and form lumps, and for reduction of the amount of used material. So far this method has worked well in preparing high-performance catalysts for ammonia cracking [105], CO_2 methanation [73,106,107], glycerol oxidation [108], and Sonogashira coupling [109].

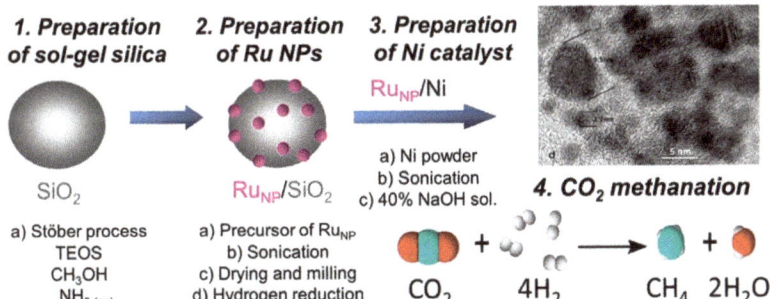

Figure 12. Preparation method of Ru/Ni catalyst for CO_2 methanation. © Adopted from [106].

4. Modeling of the Methanation Catalysis—The Determination of Research Clues

Modelling and simulations in silico are more and more often used in designing and optimizing methanation processes [68,110,111]. In such studies the kinetics of CO_2 methanation is usually modelled by a combination of CO methanation and reversed water-gas shift reaction (RWGSR) [112–114]. The resultant process depends on the rates of both these reactions. The quality of the forecasted model depends on the knowledge of reaction mechanisms and elementary stages, which determine expressions for reaction rates. However, the learning of an exact mechanism and kinetic description is not always unambiguous. This may be explained by varying reaction conditions (e.g., different values of temperatures or partial pressures), the concept of reactor and the applied catalyst, or by assumptions or the computational method (Langmuir-Hinshelwood, Power Law, elementary reactions, stages of reaction rate) [111]. However, theoretical models are necessary to design catalysts [115]. It was observed that activation energies for elementary surface reactions on catalyst are strongly correlated with adsorption energies, which facilitates identification of significant descriptors [68]. This is illustrated in Figure 13, using the example of CO methanation.

The effect of high dissociation energy is typical of a densely packed surface, while certain surface features (edges, angles, steps, and kinks) enable us to lower the energy barrier [116,117]. Therefore, an active place on the catalyst surface is identified by a convenient nucleation place. The comparison of various metallic surfaces of catalysts (Figure 13a) allows us to state that the activation barrier for CO, CH_4, and H_2O is related to the surface stability of carbon (C) and oxygen (O) forms [68]. The more stable these atoms are, the lower the CO and CH_4 dissociation barrier, and the higher the H_2O formation barrier. It was found that the activation energies also essentially depend linearly on the reaction energy acc. to the so-called Brønsted-Evans-Polanyi relationship (BEP) (Figure 13b) [118]. This enables us to make the rate of reaction on metal surfaces of various catalysts directly dependent on the CO dissociation energy (Figure 13c) [119]. In the case of poor adsorption (right part of graph in Figure 13c), the barrier for product dissociation is high, which limits the reaction rate. For a strong adsorption the rate of removing the adsorbed C and O from the surface is small, hence the barrier for product formation is high. The optimum is situated between these two limits. This effect is a well-known Sabatier rule [120]. In addition, for combinations of different materials, the scaling relationships for the adsorption and energy of transition state of the reaction are unlimited and it becomes possible to optimally adjust the catalysts' activity or selectivity even in the next catalytic sequences [121,122]. Furthermore, this search for catalytic materials is currently supported by machine learning [123]. For example, a sample of a heterogeneous catalyst in a set of different catalysts—catalyst space (defined by composition, carrier type, and particle size) can be described by its features in a certain feature space that is defined by physical properties, atomic properties, and electronic structure. Then machine learning algorithms can generate models or find descriptors that map the features that describe catalysts to their figures of merit (defined by selectivity, activity, and stability). The latest research shows that, thanks to machine learning methods, it is already possible to predict catalytic activity values, reaction descriptors, and potential energy surfaces, and to screen optimal catalysts [123–125].

The designing of catalytic materials with target properties must be described by both the basic (descriptors of anticipated properties) and empirical (measured properties) data. In addition, it is important to gather the data in a structured way, and to consider the possibility of their reorganization and export to any format, so that their processing would be easy and widely available. As a team we have drawn attention to this in the paper "Functional and Material Properties in Nanocatalyst Design: A Data Handling and Sharing Problem" [126], and by creating the "Catalytic Material Database" (CMD), available at cmd.us.edu.pl. The experimental data for heterogeneous catalysts, used mainly in carbon oxides methanation, are gathered in this database. More information on this is available on the database website.

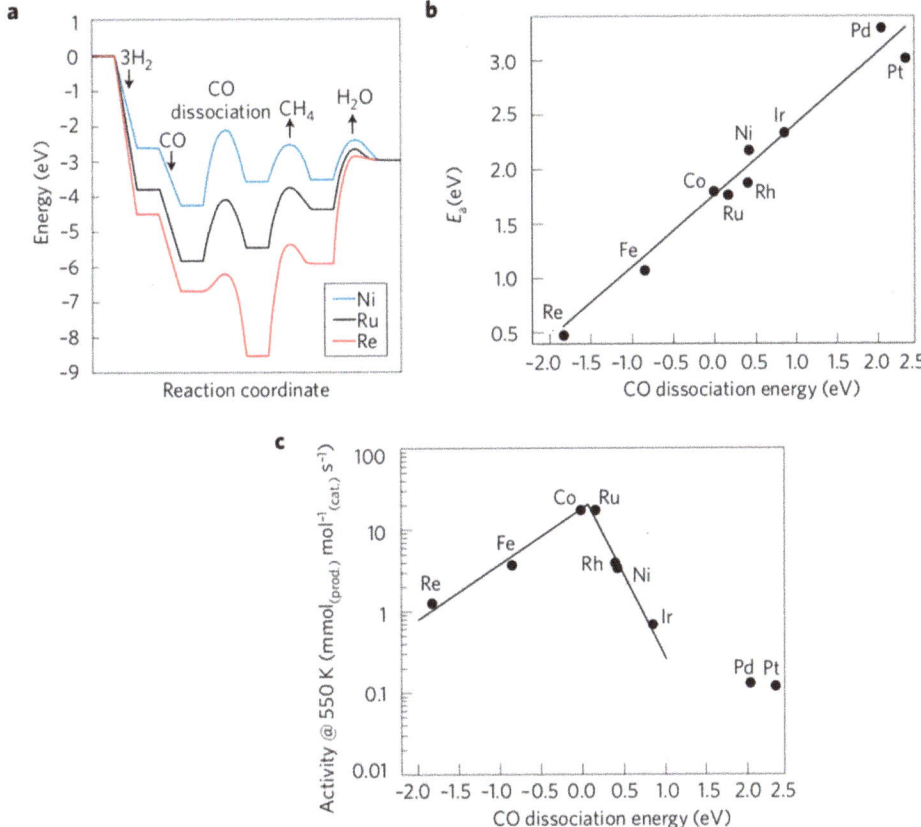

Figure 13. Identification of a descriptor for the CO methanation. © Adopted from [115,119]. (**a**) Calculated energy diagrams for CO methanation over Ni, Ru, and Re. (**b**) Brønsted–Evans–Polanyi relation for CO dissociation over transition metal surfaces. The transition state potential energy, Ea, is linearly related to the CO dissociation energy. (**c**) The corresponding measured volcano-relation for the methanation rate.

Author Contributions: Conceptualization, D.L.; methodology, D.L. and J.P.; validation, D.L., J.P. and M.K.; formal analysis, D.L. and M.K.; investigation, D.L., J.P. and M.K.; resources, J.P.; data curation, D.L., J.P. and M.K.; writing—original draft preparation, D.L. and J.P.; writing—review and editing, D.L., J.P. and M.K.; visualization, D.L. and M.K.; supervision, J.P. and D.L.; project administration, D.L.; funding acquisition, J.P. All authors have read and agreed to the published version of the manuscript.

Funding: This research was funded by National Science Center OPUS 2018/29/B/ST8/02303.

Institutional Review Board Statement: Not applicable.

Informed Consent Statement: Not applicable.

Data Availability Statement: Not applicable.

Acknowledgments: Jaroslaw Polanski would like to acknowledge Zielony Horyzont: New Energy project ZFIN 40001022 for support.

Conflicts of Interest: The authors declare no conflict of interest. The funders had no role in the design of the study; in the collection, analyses, or interpretation of data; in the writing of the manuscript, or in the decision to publish the results.

References

1. CO$_2$ Annual Mean Data—NOAA Data. Available online: Ftp://aftp.cmdl.noaa.gov/products/trends/co2/co2_annmean_mlo.txt (accessed on 20 August 2021).
2. Brodziński, Z.; Kramarz, M.; Sławomirski, M.R. *Energia Odnawialna Wizytówką Nowoczesnej Gospodarki*; Wydawnictwo Adam Marszałek: Toruń, Poland, 2016; ISBN 978-83-8019-509-7.
3. Smith, M.R.; Myers, S.S. Impact of Anthropogenic CO$_2$ Emissions on Global Human Nutrition. *Nat. Clim. Chang.* **2018**, *8*, 834–839. [CrossRef]
4. Fletcher, S.E.M. Ocean Circulation Drove Increase in CO$_2$ Uptake. *Nature* **2017**, *542*, 169–170. [CrossRef] [PubMed]
5. DeVries, T.; Holzer, M.; Primeau, F. Recent Increase in Oceanic Carbon Uptake Driven by Weaker Upper-Ocean Overturning. *Nature* **2017**, *542*, 215–218. [CrossRef] [PubMed]
6. Penuelas, J.; Fernández-Martínez, M.; Vallicrosa, H.; Maspons, J.; Zuccarini, P.; Carnicer, J.; Sanders, T.G.M.; Krüger, I.; Obersteiner, M.; Janssens, I.A.; et al. Increasing Atmospheric CO$_2$ Concentrations Correlate with Declining Nutritional Status of European Forests. *Commun. Biol.* **2020**, *3*, 125. [CrossRef]
7. European Union Emissions Trading System, EU ETS. Available online: https://ec.europa.eu/clima/policies/ets_en (accessed on 20 August 2021).
8. Pre-2020 Ambition and Implementation-CO$_2$ Reduction-Kyoto Protocol. Available online: https://unfccc.int/topics/pre-2020 (accessed on 20 August 2021).
9. Kyoto Protocol—Reference Manual. Available online: https://unfccc.int/sites/default/files/08_unfccc_kp_ref_manual.pdf (accessed on 20 August 2020).
10. De Falco, M.D.; Iaquaniello, G.; Centi, G. (Eds.) *CO$_2$: A Valuable Source of Carbon*, 1st ed.; Green Energy and Technology; Springer: London, UK, 2013; ISBN 978-1-4471-5119-7.
11. 2030 Climate & Energy Framework. Available online: https://ec.europa.eu/clima/policies/strategies/2030_en (accessed on 20 August 2021).
12. Crippa, M.; Oreggioni, G.; Guizzardi, D.; Muntean, M.; Schaaf, E.; Lo Vullo, E.; Solazzo, E.; Monforti-Ferrario, F.; Olivier, J.G.J.; Vignati, E. *Fossil CO$_2$ and GHG Emissions of All World Countries: 2019 Report*; Publications Office of the European Union: Luxemburg, 2019; ISBN 978-92-76-11100-9.
13. Infografika: Emisje Gazów Cieplarnianych w Unii Europejskiej. Available online: https://www.europarl.europa.eu/news/pl/headlines/society/20180301STO98928/infografika-emisje-gazow-cieplarnianych-w-unii-europejskiej (accessed on 20 August 2021).
14. Global Greenhouse Gas Emissions Data. Available online: https://www.epa.gov/ghgemissions/global-greenhouse-gas-emissions-data (accessed on 20 August 2021).
15. Rozporządzenie Parlamentu Europejskiego i Rady (UE) w Sprawie Wiążących Rocznych Redukcji Emisji Gazów Cieplarnianych. Available online: https://eur-lex.europa.eu/legal-content/PL/TXT/?qid=1582275556293&uri=CELEX:32018R0842 (accessed on 20 August 2021).
16. Quadrelli, E.A.; Centi, G.; Duplan, J.-L.; Perathoner, S. Carbon Dioxide Recycling: Emerging Large-Scale Technologies with Industrial Potential. *ChemSusChem* **2011**, *4*, 1194–1215. [CrossRef]
17. Aresta, M. (Ed.) *Carbon Dioxide as Chemical Feedstock*, 1st ed.; Wiley: Weinheim, Germany, 2010; ISBN 978-3-527-32475-0.
18. Mikkelsen, M.; Jørgensen, M.; Krebs, F.C. The Teraton Challenge. A Review of Fixation and Transformation of Carbon Dioxide. *Energy Environ. Sci.* **2010**, *3*, 43–81. [CrossRef]
19. Peters, M.; Köhler, B.; Kuckshinrichs, W.; Leitner, W.; Markewitz, P.; Müller, T.E. Chemical Technologies for Exploiting and Recycling Carbon Dioxide into the Value Chain. *ChemSusChem* **2011**, *4*, 1216–1240. [CrossRef]
20. Centi, G.; Perathoner, S. Opportunities and Prospects in the Chemical Recycling of Carbon Dioxide to Fuels. *Catal. Today* **2009**, *148*, 191–205. [CrossRef]
21. Dorner, R.W.; Hardy, D.R.; Williams, F.W.; Willauer, H.D. Heterogeneous Catalytic CO$_2$ Conversion to Value-Added Hydrocarbons. *Energy Environ. Sci.* **2010**, *3*, 884. [CrossRef]
22. International Energy Agency; United Nations Industrial Development Organization. *Carbon Capture and Storage in Industrial Applications*; IEA Technology Roadmaps; OECD: Paris, France, 2012; ISBN 978-92-64-13066-1.
23. Aziz, M.A.A.; Jalil, A.A.; Triwahyono, S.; Ahmad, A. CO$_2$ Methanation over Heterogeneous Catalysts: Recent Progress and Future Prospects. *Green Chem.* **2015**, *17*, 2647–2663. [CrossRef]
24. Carbon Dioxide Emission by Source Sector (Source: EEA). Available online: https://appsso.eurostat.ec.europa.eu/nui/submitViewTableAction.do (accessed on 20 August 2021).
25. Triantafyllidis, K.S.; Lappas, A.A.; Stöcker, M. (Eds.) *The Role of Catalysis for the Sustainable Production of Bio-Fuels and Bio-Chemicals*, 1st ed.; Elsevier: Amsterdam, The Netherlands; Boston, MA, USA, 2013; ISBN 978-0-444-56330-9.
26. Energy Efficiency and Renewable Energy. *Energy-Intensive Processes Portfolio: Addressing Key Energy Challenges Across U.S. Industry*; Industrial Technologies Program; U.S. Department of Energy: Washington, DC, USA, 2011.
27. Metz, B.; Intergovernmental Panel on Climate Change (Eds.) *IPCC Special Report on Carbon Dioxide Capture and Storage*; Cambridge University Press, for the Intergovernmental Panel on Climate Change: Cambridge, UK, 2005; ISBN 978-0-521-86643-9.
28. Wulf, C.; Linßen, J.; Zapp, P. Review of Power-to-Gas Projects in Europe. *Energy Procedia* **2018**, *155*, 367–378. [CrossRef]

29. Sterner, M. *Bioenergy and Renewable Power Methane in Integrated 100% Renewable Energy Systems: Limiting Global Warming by Transforming Energy Systems*; Erneuerbare Energien und Energieeffizienz/Renewable Energies and Energy Efficiency; Kassel University Press: Kassel, Germany, 2010; ISBN 978-3-89958-798-2.
30. Olah, G.A.; Goeppert, A.; Prakash, G.K.S. *Beyond Oil and Gas: The Methanol Economy*; Wiley-VCH: Weinheim, Germany, 2011; ISBN 978-3-527-64463-6.
31. Grignard, B.; Gennen, S.; Jérôme, C.; Kleij, A.W.; Detrembleur, C. Advances in the Use of CO_2 as a Renewable Feedstock for the Synthesis of Polymers. *Chem. Soc. Rev.* **2019**, *48*, 4466–4514. [CrossRef]
32. Pescarmona, P.P. Cyclic Carbonates Synthesised from CO_2: Applications, Challenges and Recent Research Trends. *Curr. Opin. Green Sustain. Chem.* **2021**, *29*, 100457. [CrossRef]
33. Kelly, M.J.; Barthel, A.; Maheu, C.; Sodpiban, O.; Dega, F.-B.; Vummaleti, S.V.C.; Abou-Hamad, E.; Pelletier, J.D.A.; Cavallo, L.; D'Elia, V.; et al. Conversion of Actual Flue Gas CO_2 via Cycloaddition to Propylene Oxide Catalyzed by a Single-Site, Recyclable Zirconium Catalyst. *J. CO_2 Util.* **2017**, *20*, 243–252. [CrossRef]
34. Sodpiban, O.; Phungpanya, C.; Del Gobbo, S.; Arayachukiat, S.; Piromchart, T.; D'Elia, V. Rational Engineering of Single-Component Heterogeneous Catalysts Based on Abundant Metal Centers for the Mild Conversion of Pure and Impure CO_2 to Cyclic Carbonates. *Chem. Eng. J.* **2021**, *422*, 129930. [CrossRef]
35. Chen, Y.; Luo, R.; Xu, Q.; Jiang, J.; Zhou, X.; Ji, H. Charged Metalloporphyrin Polymers for Cooperative Synthesis of Cyclic Carbonates from CO_2 under Ambient Conditions. *ChemSusChem* **2017**, *10*, 2534–2541. [CrossRef]
36. Fukuoka, S.; Fukawa, I.; Tojo, M.; Oonishi, K.; Hachiya, H.; Aminaka, M.; Hasegawa, K.; Komiya, K. A Novel Non-Phosgene Process for Polycarbonate Production from CO_2: Green and Sustainable Chemistry in Practice. *Catal. Surv. Asia* **2010**, *14*, 146–163. [CrossRef]
37. Demirbas, A.; Demirbas, M.F. *Algae Energy: Algae as a New Source of Biodiesel*; Green Energy and Technology; Springer: Dordrecht, The Netherlands; New York, NY, USA, 2010; ISBN 978-1-84996-050-2.
38. Chiappe, C.; Mezzetta, A.; Pomelli, C.S.; Iaquaniello, G.; Gentile, A.; Masciocchi, B. Development of Cost-Effective Biodiesel from Microalgae Using Protic Ionic Liquids. *Green Chem.* **2016**, *18*, 4982–4989. [CrossRef]
39. Antonovsky, N.; Gleizer, S.; Noor, E.; Zohar, Y.; Herz, E.; Barenholz, U.; Zelcbuch, L.; Amram, S.; Wides, A.; Tepper, N.; et al. Sugar Synthesis from CO_2 in Escherichia Coli. *Cell* **2016**, *166*, 115–125. [CrossRef]
40. Callaway, E.E. Coli Bacteria Engineered to Eat Carbon Dioxide. *Nature* **2019**, *576*, 19–20. [CrossRef]
41. Sharma, T.; Sharma, S.; Kamyab, H.; Kumar, A. Energizing the CO_2 Utilization by Chemo-Enzymatic Approaches and Potentiality of Carbonic Anhydrases: A Review. *J. Clean. Prod.* **2020**, *247*, 119138. [CrossRef]
42. Eckert, C.A.; Knutson, B.L.; Debenedetti, P.G. Supercritical Fluids as Solvents for Chemical and Materials Processing. *Nature* **1996**, *383*, 313–318. [CrossRef]
43. Sodeifian, G.; Sajadian, S.A. Solubility Measurement and Preparation of Nanoparticles of an Anticancer Drug (Letrozole) Using Rapid Expansion of Supercritical Solutions with Solid Cosolvent (RESS-SC). *J. Supercrit. Fluids* **2018**, *133*, 239–252. [CrossRef]
44. Sodeifian, G.; Sajadian, S.A.; Saadati Ardestani, N.; Razmimanesh, F. Production of Loratadine Drug Nanoparticles Using Ultrasonic-Assisted Rapid Expansion of Supercritical Solution into Aqueous Solution (US-RESSAS). *J. Supercrit. Fluids* **2019**, *147*, 241–253. [CrossRef]
45. Sodeifian, G.; Garlapati, C.; Razmimanesh, F.; Ghanaat-Ghamsari, M. Measurement and Modeling of Clemastine Fumarate (Antihistamine Drug) Solubility in Supercritical Carbon Dioxide. *Sci. Rep.* **2021**, *11*, 24344. [CrossRef]
46. Sodeifian, G.; Surya Alwi, R.; Razmimanesh, F.; Abadian, M. Solubility of Dasatinib Monohydrate (Anticancer Drug) in Supercritical CO_2: Experimental and Thermodynamic Modeling. *J. Mol. Liq.* **2022**, *346*, 117899. [CrossRef]
47. Sodeifian, G.; Sajadian, S.A.; Derakhsheshpour, R. CO_2 Utilization as a Supercritical Solvent and Supercritical Antisolvent in Production of Sertraline Hydrochloride Nanoparticles. *J. CO_2 Util.* **2022**, *55*, 101799. [CrossRef]
48. Ameri, A.; Sodeifian, G.; Sajadian, S.A. Lansoprazole Loading of Polymers by Supercritical Carbon Dioxide Impregnation: Impacts of Process Parameters. *J. Supercrit. Fluids* **2020**, *164*, 104892. [CrossRef]
49. Saadati Ardestani, N.; Sodeifian, G.; Sajadian, S.A. Preparation of Phthalocyanine Green Nano Pigment Using Supercritical CO_2 Gas Antisolvent (GAS): Experimental and Modeling. *Heliyon* **2020**, *6*, e04947. [CrossRef]
50. Sodeifian, G.; Saadati Ardestani, N.; Sajadian, S.A.; Soltani Panah, H. Experimental Measurements and Thermodynamic Modeling of Coumarin-7 Solid Solubility in Supercritical Carbon Dioxide: Production of Nanoparticles via RESS Method. *Fluid Phase Equilibria* **2019**, *483*, 122–143. [CrossRef]
51. Sodeifian, G.; Ansari, K. Optimization of Ferulago Angulata Oil Extraction with Supercritical Carbon Dioxide. *J. Supercrit. Fluids* **2011**, *57*, 38–43. [CrossRef]
52. Sodeifian, G.; Saadati Ardestani, N.; Sajadian, S.A.; Ghorbandoost, S. Application of Supercritical Carbon Dioxide to Extract Essential Oil from Cleome Coluteoides Boiss: Experimental, Response Surface and Grey Wolf Optimization Methodology. *J. Supercrit. Fluids* **2016**, *114*, 55–63. [CrossRef]
53. White, M.T.; Bianchi, G.; Chai, L.; Tassou, S.A.; Sayma, A.I. Review of Supercritical CO_2 Technologies and Systems for Power Generation. *Appl. Therm. Eng.* **2021**, *185*, 116447. [CrossRef]
54. Climeworks-CO_2 Removal. Available online: https://climeworks.com (accessed on 23 August 2021).
55. Ghiasi, M.; Zeinali, P.; Gholami, S.; Zahedi, M. Separation of CH_4, H_2S, N_2 and CO_2 Gases Using Four Types of Nanoporous Graphene Cluster Model: A Quantum Chemical Investigation. *J. Mol. Model.* **2021**, *27*, 201. [CrossRef]

56. Shaikh, A.R.; Ashraf, M.; AlMayef, T.; Chawla, M.; Poater, A.; Cavallo, L. Amino Acid Ionic Liquids as Potential Candidates for CO_2 Capture: Combined Density Functional Theory and Molecular Dynamics Simulations. *Chem. Phys. Lett.* **2020**, *745*, 137239. [CrossRef]
57. Sabatier, P.; Senderens, J.-B. Nouvelles Synthèses Du Méthane. *Comptes Rendus l'Académie Sci.* **1902**, *134*, 514–516.
58. Barbarossa, V.; Vanga, G. *Energia, Ambiente e Innovazione*; ENEA: Rome, Italy, 2011; pp. 82–85.
59. Gao, J.; Liu, Q.; Gu, F.; Liu, B.; Zhong, Z.; Su, F. Recent Advances in Methanation Catalysts for the Production of Synthetic Natural Gas. *RSC Adv.* **2015**, *5*, 22759–22776. [CrossRef]
60. Artz, J.; Müller, T.E.; Thenert, K.; Kleinekorte, J.; Meys, R.; Sternberg, A.; Bardow, A.; Leitner, W. Sustainable Conversion of Carbon Dioxide: An Integrated Review of Catalysis and Life Cycle Assessment. *Chem. Rev.* **2018**, *118*, 434–504. [CrossRef]
61. Centi, G.; Perathoner, S. Heterogeneous Catalytic Reactions with CO_2: Status and Perspectives. In *Studies in Surface Science and Catalysis*; Elsevier: Amsterdam, The Netherlands, 2004; Volume 153, pp. 1–8, ISBN 978-0-444-51600-8.
62. Jessop, P.G.; Joó, F.; Tai, C.-C. Recent Advances in the Homogeneous Hydrogenation of Carbon Dioxide. *Coord. Chem. Rev.* **2004**, *248*, 2425–2442. [CrossRef]
63. Rönsch, S.; Schneider, J.; Matthischke, S.; Schlüter, M.; Götz, M.; Lefebvre, J.; Prabhakaran, P.; Bajohr, S. Review on Methanation—From Fundamentals to Current Projects. *Fuel* **2016**, *166*, 276–296. [CrossRef]
64. Choe, S.J.; Kang, H.-J.; Kim, S.-J.; Park, S.-B.; Park, D.H.; Huh, D.S. Adsorbed Carbon Formation and Carbon Hydrogenation for CO_2 Methanation on the Ni(111) Surface: ASED-MO Study. *Bull. Korean Chem. Soc.* **2005**, *26*, 1682–1688. [CrossRef]
65. Hwang, S.; Lee, J.; Hong, U.G.; Seo, J.G.; Jung, J.C.; Koh, D.J.; Lim, H.; Byun, C.; Song, I.K. Methane Production from Carbon Monoxide and Hydrogen over Nickel–Alumina Xerogel Catalyst: Effect of Nickel Content. *J. Ind. Eng. Chem.* **2011**, *17*, 154–157. [CrossRef]
66. Tada, S.; Ikeda, S.; Shimoda, N.; Honma, T.; Takahashi, M.; Nariyuki, A.; Satokawa, S. Sponge Ni Catalyst with High Activity in CO_2 Methanation. *Int. J. Hydrogen Energy* **2017**, *42*, 30126–30134. [CrossRef]
67. Polanski, J.; Lach, D.; Kapkowski, M.; Bartczak, P.; Siudyga, T.; Smolinski, A. Ru and Ni—Privileged Metal Combination for Environmental Nanocatalysis. *Catalysts* **2020**, *10*, 992. [CrossRef]
68. Bligaard, T.; Nørskov, J.K.; Dahl, S.; Matthiesen, J.; Christensen, C.H.; Sehested, J. The Brønsted–Evans–Polanyi Relation and the Volcano Curve in Heterogeneous Catalysis. *J. Catal.* **2004**, *224*, 206–217. [CrossRef]
69. Kai, T.; Yamasaki, Y.; Takahashi, T.; Masumoto, T.; Kimura, H. Increase in the Thermal Stability during the Methanation of CO_2 over a Rh Catalyst Prepared from an Amorphous Alloy. *Can. J. Chem. Eng.* **1998**, *76*, 331–335. [CrossRef]
70. Martin, N.M.; Hemmingsson, F.; Wang, X.; Merte, L.R.; Hejral, U.; Gustafson, J.; Skoglundh, M.; Meira, D.M.; Dippel, A.-C.; Gutowski, O.; et al. Structure–Function Relationship during CO_2 Methanation over Rh/Al_2O_3 and Rh/SiO_2 Catalysts under Atmospheric Pressure Conditions. *Catal. Sci. Technol.* **2018**, *8*, 2686–2696. [CrossRef]
71. Benítez, J.J.; Alvero, R.; Capitán, M.J.; Carrizosa, I.; Odriozola, J.A. DRIFTS Study of Adsorbed Formate Species in the Carbon Dioxide and Hydrogen Reaction over Rhodium Catalysts. *Appl. Catal.* **1991**, *71*, 219–231. [CrossRef]
72. Abe, T.; Tanizawa, M.; Watanabe, K.; Taguchi, A. CO_2 Methanation Property of Ru Nanoparticle-Loaded TiO_2 Prepared by a Polygonal Barrel-Sputtering Method. *Energy Environ. Sci.* **2009**, *2*, 315. [CrossRef]
73. Siudyga, T.; Kapkowski, M.; Janas, D.; Wasiak, T.; Sitko, R.; Zubko, M.; Szade, J.; Balin, K.; Klimontko, J.; Lach, D.; et al. Nano-Ru Supported on Ni Nanowires for Low-Temperature Carbon Dioxide Methanation. *Catalysts* **2020**, *10*, 513. [CrossRef]
74. Snytnikov, P.V.; Zyryanova, M.M.; Sobyanin, V.A. CO-Cleanup of Hydrogen-Rich Stream for LT PEM FC Feeding: Catalysts and Their Performance in Selective CO Methanation. *Top. Catal.* **2016**, *59*, 1394–1412. [CrossRef]
75. Ross, J.R.H. *Contemporary Catalysis: Fundamentals and Current Applications*; Elsevier: Amsterdam, The Netherlands, 2019; ISBN 978-0-444-63474-0.
76. Martínez, J.; Hernández, E.; Alfaro, S.; López Medina, R.; Valverde Aguilar, G.; Albiter, E.; Valenzuela, M. High Selectivity and Stability of Nickel Catalysts for CO_2 Methanation: Support Effects. *Catalysts* **2018**, *9*, 24. [CrossRef]
77. Le, T.A.; Kim, M.S.; Lee, S.H.; Kim, T.W.; Park, E.D. CO and CO_2 Methanation over Supported Ni Catalysts. *Catal. Today* **2017**, *293–294*, 89–96. [CrossRef]
78. Fan, M.-T.; Miao, K.-P.; Lin, J.-D.; Zhang, H.-B.; Liao, D.-W. Mg-Al Oxide Supported Ni Catalysts with Enhanced Stability for Efficient Synthetic Natural Gas from Syngas. *Appl. Surf. Sci.* **2014**, *307*, 682–688. [CrossRef]
79. Hu, D.; Gao, J.; Ping, Y.; Jia, L.; Gunawan, P.; Zhong, Z.; Xu, G.; Gu, F.; Su, F. Enhanced Investigation of CO Methanation over Ni/Al_2O_3 Catalysts for Synthetic Natural Gas Production. *Ind. Eng. Chem. Res.* **2012**, *51*, 4875–4886. [CrossRef]
80. Qin, H.; Guo, C.; Wu, Y.; Zhang, J. Effect of La_2O_3 Promoter on NiO/Al_2O_3 Catalyst in CO Methanation. *Korean J. Chem. Eng.* **2014**, *31*, 1168–1173. [CrossRef]
81. Liu, Q.; Gu, F.; Lu, X.; Liu, Y.; Li, H.; Zhong, Z.; Xu, G.; Su, F. Enhanced Catalytic Performances of Ni/Al_2O_3 Catalyst via Addition of V_2O_3 for CO Methanation. *Appl. Catal. A-Gen.* **2014**, *488*, 37–47. [CrossRef]
82. Liu, H.; Zou, X.; Wang, X.; Lu, X.; Ding, W. Effect of CeO_2 Addition on Ni/Al_2O_3 Catalysts for Methanation of Carbon Dioxide with Hydrogen. *J. Nat. Gas Chem.* **2012**, *21*, 703–707. [CrossRef]
83. Campbell, C.T.; Goodman, D.W. A Surface Science Investigation of the Role of Potassium Promoters in Nickel Catalysts for CO Hydrogenation. *Surf. Sci.* **1982**, *123*, 413–426. [CrossRef]

84. Aldana, P.A.U.; Ocampo, F.; Kobl, K.; Louis, B.; Thibault-Starzyk, F.; Daturi, M.; Bazin, P.; Thomas, S.; Roger, A.C. Catalytic CO_2 Valorization into CH_4 on Ni-Based Ceria-Zirconia. Reaction Mechanism by Operando IR Spectroscopy. *Catal. Today* **2013**, *215*, 201–207. [CrossRef]
85. Pan, Q.; Peng, J.; Wang, S.; Wang, S. In Situ FTIR Spectroscopic Study of the CO_2 Methanation Mechanism on $Ni/Ce_{0.5}Zr_{0.5}O_2$. *Catal. Sci. Technol.* **2014**, *4*, 502–509. [CrossRef]
86. Fujita, S.; Nakamura, M.; Doi, T.; Takezawa, N. Mechanisms of Methanation of Carbon Dioxide and Carbon Monoxide over Nickel/Alumina Catalysts. *Appl. Catal. A-Gen.* **1993**, *104*, 87–100. [CrossRef]
87. Karelovic, A.; Ruiz, P. Improving the Hydrogenation Function of $Pd/\gamma-Al_2O_3$ Catalyst by $Rh/\gamma-Al_2O_3$ Addition in CO_2 Methanation at Low Temperature. *ACS Catal.* **2013**, *3*, 2799–2812. [CrossRef]
88. Aziz, M.A.A.; Jalil, A.A.; Triwahyono, S.; Sidik, S.M. Methanation of Carbon Dioxide on Metal-Promoted Mesostructured Silica Nanoparticles. *Appl. Catal. A-Gen.* **2014**, *486*, 115–122. [CrossRef]
89. Aziz, M.A.A.; Jalil, A.A.; Triwahyono, S.; Mukti, R.R.; Taufiq-Yap, Y.H.; Sazegar, M.R. Highly Active Ni-Promoted Mesostructured Silica Nanoparticles for CO_2 Methanation. *Appl. Catal. B* **2014**, *147*, 359–368. [CrossRef]
90. Aziz, M.A.A.; Jalil, A.A.; Triwahyono, S.; Saad, M.W.A. CO_2 Methanation over Ni-Promoted Mesostructured Silica Nanoparticles: Influence of Ni Loading and Water Vapor on Activity and Response Surface Methodology Studies. *Chem. Eng. J.* **2015**, *260*, 757–764. [CrossRef]
91. Kärger, J.; Ruthven, D.M.; Theodorou, D.N. *Diffusion in Nanoporous Materials*, 1st ed.; Wiley: Hoboken, NJ, USA, 2012; ISBN 978-3-527-31024-1.
92. Kärger, J.; Goepel, M.; Gläser, R. Diffusion in Nanocatalysis. In *Nanotechnology in Catalysis*; Van de Voorde, M., Sels, B., Eds.; Wiley-VCH Verlag GmbH & Co. KGaA: Weinheim, Germany, 2017; pp. 293–334, ISBN 978-3-527-69982-7.
93. Tesser, R.; Santacesaria, E. Revisiting the Role of Mass and Heat Transfer in Gas–Solid Catalytic Reactions. *Processes* **2020**, *8*, 1599. [CrossRef]
94. Dai, X.; Sun, Y. Reduction of Carbon Dioxide on Photoexcited Nanoparticles of VIII Group Metals. *Nanoscale* **2019**, *11*, 16723–16732. [CrossRef]
95. Robatjazi, H.; Zhao, H.; Swearer, D.F.; Hogan, N.J.; Zhou, L.; Alabastri, A.; McClain, M.J.; Nordlander, P.; Halas, N.J. Plasmon-Induced Selective Carbon Dioxide Conversion on Earth-Abundant Aluminum-Cuprous Oxide Antenna-Reactor Nanoparticles. *Nat. Commun.* **2017**, *8*, 27. [CrossRef]
96. Kale, M.J.; Avanesian, T.; Xin, H.; Yan, J.; Christopher, P. Controlling Catalytic Selectivity on Metal Nanoparticles by Direct Photoexcitation of Adsorbate–Metal Bonds. *Nano Lett.* **2014**, *14*, 5405–5412. [CrossRef]
97. Kim, Y.; Dumett Torres, D.; Jain, P.K. Activation Energies of Plasmonic Catalysts. *Nano Lett.* **2016**, *16*, 3399–3407. [CrossRef]
98. Pinchuk, A.; von Plessen, G.; Kreibig, U. Influence of Interband Electronic Transitions on the Optical Absorption in Metallic Nanoparticles. *J. Phys. D Appl. Phys.* **2004**, *37*, 3133–3139. [CrossRef]
99. Pinchuk, A.; Kreibig, U.; Hilger, A. Optical Properties of Metallic Nanoparticles: Influence of Interface Effects and Interband Transitions. *Surf. Sci.* **2004**, *557*, 269–280. [CrossRef]
100. Zhang, C.; Kong, T.; Fu, Z.; Zhang, Z.; Zheng, H. Hot Electron and Thermal Effects in Plasmonic Catalysis of Nanocrystal Transformation. *Nanoscale* **2020**, *12*, 8768–8774. [CrossRef] [PubMed]
101. Kim, C.; Hyeon, S.; Lee, J.; Kim, W.D.; Lee, D.C.; Kim, J.; Lee, H. Energy-Efficient CO_2 Hydrogenation with Fast Response Using Photoexcitation of CO_2 Adsorbed on Metal Catalysts. *Nat. Commun.* **2018**, *9*, 3027. [CrossRef] [PubMed]
102. Schwarz, J.A.; Contescu, C.; Contescu, A. Methods for Preparation of Catalytic Materials. *Chem. Rev.* **1995**, *95*, 477–510. [CrossRef]
103. Stöber, W.; Fink, A.; Bohn, E. Controlled Growth of Monodisperse Silica Spheres in the Micron Size Range. *J. Colloid Interface Sci.* **1968**, *26*, 62–69. [CrossRef]
104. Rao, K.S.; El-Hami, K.; Kodaki, T.; Matsushige, K.; Makino, K. A Novel Method for Synthesis of Silica Nanoparticles. *J. Colloid Interface Sci.* **2005**, *289*, 125–131. [CrossRef]
105. Polanski, J.; Bartczak, P.; Ambrozkiewicz, W.; Sitko, R.; Siudyga, T.; Mianowski, A.; Szade, J.; Balin, K.; Lelątko, J. Ni-Supported Pd Nanoparticles with Ca Promoter: A New Catalyst for Low-Temperature Ammonia Cracking. *PLoS ONE* **2015**, *10*, e0136805. [CrossRef]
106. Polanski, J.; Siudyga, T.; Bartczak, P.; Kapkowski, M.; Ambrozkiewicz, W.; Nobis, A.; Sitko, R.; Klimontko, J.; Szade, J.; Lelątko, J. Oxide Passivated Ni-Supported Ru Nanoparticles in Silica: A New Catalyst for Low-Temperature Carbon Dioxide Methanation. *Appl. Catal. B.* **2017**, *206*, 16–23. [CrossRef]
107. Siudyga, T.; Kapkowski, M.; Bartczak, P.; Zubko, M.; Szade, J.; Balin, K.; Antoniotti, S.; Polanski, J. Ultra-Low Temperature Carbon (Di)Oxide Hydrogenation Catalyzed by Hybrid Ruthenium–Nickel Nanocatalysts: Towards Sustainable Methane Production. *Green Chem.* **2020**, *22*, 5143–5150. [CrossRef]
108. Kapkowski, M.; Bartczak, P.; Korzec, M.; Sitko, R.; Szade, J.; Balin, K.; Lelątko, J.; Polanski, J. SiO_2-, Cu-, and Ni-Supported Au Nanoparticles for Selective Glycerol Oxidation in the Liquid Phase. *J. Catal.* **2014**, *319*, 110–118. [CrossRef]
109. Korzec, M.; Bartczak, P.; Niemczyk, A.; Szade, J.; Kapkowski, M.; Zenderowska, P.; Balin, K.; Lelątko, J.; Polanski, J. Bimetallic Nano-Pd/PdO/Cu System as a Highly Effective Catalyst for the Sonogashira Reaction. *J. Catal.* **2014**, *313*, 1–8. [CrossRef]
110. Rönsch, S.; Köchermann, J.; Schneider, J.; Matthischke, S. Global Reaction Kinetics of CO and CO_2 Methanation for Dynamic Process Modeling. *Chem. Eng. Technol.* **2016**, *39*, 208–218. [CrossRef]

111. Hernandez Lalinde, J.A.; Roongruangsree, P.; Ilsemann, J.; Bäumer, M.; Kopyscinski, J. CO_2 Methanation and Reverse Water Gas Shift Reaction. Kinetic Study Based on in Situ Spatially-Resolved Measurements. *Chem. Eng. J.* **2020**, *390*, 124629. [CrossRef]
112. Kopyscinski, J.; Schildhauer, T.J.; Biollaz, S.M.A. Methanation in a Fluidized Bed Reactor with High Initial CO Partial Pressure: Part II— Modeling and Sensitivity Study. *Chem. Eng. Sci.* **2011**, *66*, 1612–1621. [CrossRef]
113. Champon, I.; Bengaouer, A.; Chaise, A.; Thomas, S.; Roger, A.-C. Carbon Dioxide Methanation Kinetic Model on a Commercial Ni/Al_2O_3 Catalyst. *J. CO_2 Util.* **2019**, *34*, 256–265. [CrossRef]
114. Xu, J.; Froment, G.F. Methane Steam Reforming, Methanation and Water-Gas Shift: I. Intrinsic Kinetics. *AIChE J.* **1989**, *35*, 88–96. [CrossRef]
115. Nørskov, J.K.; Bligaard, T.; Rossmeisl, J.; Christensen, C.H. Towards the Computational Design of Solid Catalysts. *Nat. Chem.* **2009**, *1*, 37–46. [CrossRef]
116. Nørskov, J.K.; Bligaard, T.; Logadottir, A.; Bahn, S.; Hansen, L.B.; Bollinger, M.; Bengaard, H.; Hammer, B.; Sljivancanin, Z.; Mavrikakis, M.; et al. Universality in Heterogeneous Catalysis. *J. Catal.* **2002**, *209*, 275–278. [CrossRef]
117. Ciobica, I.M.; van Santen, R.A. Carbon Monoxide Dissociation on Planar and Stepped Ru(0001) Surfaces. *J. Phys. Chem. B* **2003**, *107*, 3808–3812. [CrossRef]
118. Michaelides, A.; Liu, Z.-P.; Zhang, C.J.; Alavi, A.; King, D.A.; Hu, P. Identification of General Linear Relationships between Activation Energies and Enthalpy Changes for Dissociation Reactions at Surfaces. *J. Am. Chem. Soc.* **2003**, *125*, 3704–3705. [CrossRef]
119. Andersson, M.; Bligaard, T.; Kustov, A.; Larsen, K.; Greeley, J.; Johannessen, T.; Christensen, C.; Norskov, J. Toward Computational Screening in Heterogeneous Catalysis: Pareto-Optimal Methanation Catalysts. *J. Catal.* **2006**, *239*, 501–506. [CrossRef]
120. Sabatier, P. Hydrogénations et déshydrogénations par catalyse. *Ber. Dtsch. Chem. Ges.* **1911**, *44*, 1984–2001. [CrossRef]
121. Kumar, G.; Nikolla, E.; Linic, S.; Medlin, J.W.; Janik, M.J. Multicomponent Catalysts: Limitations and Prospects. *ACS Catal.* **2018**, *8*, 3202–3208. [CrossRef]
122. Greeley, J. Theoretical Heterogeneous Catalysis: Scaling Relationships and Computational Catalyst Design. *Annu. Rev. Chem. Biomol. Eng.* **2016**, *7*, 605–635. [CrossRef]
123. Goldsmith, B.R.; Esterhuizen, J.; Liu, J.; Bartel, C.J.; Sutton, C. Machine Learning for Heterogeneous Catalyst Design and Discovery. *AIChE J.* **2018**, *64*, 2311–2323. [CrossRef]
124. Suzuki, K.; Toyao, T.; Maeno, Z.; Takakusagi, S.; Shimizu, K.; Takigawa, I. Statistical Analysis and Discovery of Heterogeneous Catalysts Based on Machine Learning from Diverse Published Data. *ChemCatChem* **2019**, *11*, 4537–4547. [CrossRef]
125. Ouyang, R.; Xie, Y.; Jiang, D. Global Minimization of Gold Clusters by Combining Neural Network Potentials and the Basin-Hopping Method. *Nanoscale* **2015**, *7*, 14817–14821. [CrossRef]
126. Lach, D.; Zhdan, U.; Smolinski, A.; Polanski, J. Functional and Material Properties in Nanocatalyst Design: A Data Handling and Sharing Problem. *IJMS* **2021**, *22*, 5176. [CrossRef]

Article

Catalytic Removal of NOx on Ceramic Foam-Supported ZnO and TiO$_2$ Nanorods Ornamented with W and V Oxides

Maciej Kapkowski [1], Tomasz Siudyga [1], Piotr Bartczak [1], Maciej Zubko [2,3], Rafal Sitko [1], Jacek Szade [4], Katarzyna Balin [4], Bartłomiej S. Witkowski [5], Monika Ożga [5], Rafał Pietruszka [5], Marek Godlewski [5] and Jaroslaw Polanski [1,*]

[1] Faculty of Science and Technology, Institute of Chemistry, University of Silesia, Szkolna 9, 40-006 Katowice, Poland; maciej.kapkowski@us.edu.pl (M.K.); tomasz.siudyga@us.edu.pl (T.S.); piotr.bartczak@us.edu.pl (P.B.); rafal.sitko@us.edu.pl (R.S.)
[2] Faculty of Science and Technology, Institute of Materials Engineering, University of Silesia, 75 Pułku Piechoty 1A, 41-500 Chorzów, Poland; maciej.zubko@us.edu.pl
[3] Department of Physics, University of Hradec Králové, Rokitanského 62, 500-03 Hradec Králové, Czech Republic
[4] Faculty of Science and Technology, August Chełkowski Institute of Physics, University of Silesia, 75 Pułku Piechoty 1A, 41-500 Chorzów, Poland; jacek.szade@us.edu.pl (J.S.); katarzyna.balin@us.edu.pl (K.B.)
[5] Institute of Physics, Polish Academy of Sciences, Al. Lotników 32/46, 02668 Warsaw, Poland; bwitkow@ifpan.edu.pl (B.S.W.); ozga@ifpan.edu.pl (M.O.); pietruszka@ifpan.edu.pl (R.P.); godlew@ifpan.edu.pl (M.G.)
* Correspondence: polanski@us.edu.pl; Tel.: +48-32-2599978

Abstract: Energy consumption steadily increases and energy production is associated with many environmental risks, e.g., generating the largest share of greenhouse gas emissions. The primary gas pollution concern is CO_2, CH_4, and nitrogen oxides (NOx). Environmental catalysis plays a pivotal role in NOx mitigation (DeNOx). This study investigated, for the first time, a collection of ceramic foams as potential catalyst support for selective catalytic NOx reduction (SCR). Ceramic foams could be an attractive support option for NOx removal. However, we should functionalize the surface of raw foams for such applications. A library of ceramic SiC, Al_2O_3, and ZrO_2 foams ornamented with nanorod ZnO and TiO$_2$ as W and V oxide support was obtained for the first time. We characterized the surface layer coating structure using the XPS, XRF and SEM, and TEM microscopy to optimize the W to V molar ratio and examine NO$_2$ mitigation as the SCR model, which was tested only very rarely. Comparing TiO$_2$ and ZnO systems reveals that the SCR conversion on ZnO appeared superior vs. the conversion on TiO$_2$, while the SiC-supported catalysts were less efficient than Al_2O_3 and ZrO_2-supported catalysts. The energy bands in optical spectra correlate with the observed activity rank.

Keywords: ceramic foams; ZnO nanorods; TiO$_2$ nanorods; NOx mitigation (deNOx); environmental nanocatalysis; selective catalytic reduction SCR; W and V catalytic sites

1. Introduction

Energy demand growth is a fundamental problem of civilization in the Anthropocene. The production of energy is, however, associated with many environmental risks. Mainly, NOx formation is a consequence of energy production. One of the targets of environmental catalysis is the mitigation of NOx from the air. The design and development of porous functional materials is an essential environmental catalysis domain [1]. In particular, porous ceramics are essential catalyst supports in this area [2]. Ceramic foams are monolithic three-dimensional structures with an 80–90% void spaces fraction. However, these materials were developed initially to filter out molten metal impurities [3], which means that their surface area is generally too low for catalytic applications. Therefore, ceramic foams are modified to increase their surface area [4]. A variety of novel catalytic applications of ceramic foams involve, for example, the catalytic pyrolysis of waste oils to renewable fuels [5], the water gas shift reaction [6], and the solar photocatalytic ozonation in water treatment using supported TiO_2 [7] methane steam reforming [8]. The advantages of the innovatively structured foam catalysts involve fluid dynamics and heat transfer phenomena, which can positively influence catalyst performance.

This study investigated a collection of ceramic foams as potential catalysts for selective catalytic NOx reduction (SCR) reactions. Despite potential advantages, the literature rarely describes the application of ceramic foams in SCR catalysis [9–11]. The availability of the commercial honeycomb or plate SCR catalysts may be one reason for this fact. NOx generation, pollution, and reduction are complex problems. First, combustion in air yields two forms of NOx, namely NO and NO_2, in the ratio NO/NOx of 0.90 to 0.95 [12]. However, the dominating NOx form in the atmosphere is NO_2 resulting from NO oxidation. Accordingly, the main issue in the environmental catalysis of NOx refers to NO, while the main topics of NOx ecotoxicology refer to NO_2. The current technology routinely uses Selective Catalytic Reduction (SCR) for NOx removal from flue gases [12,13]. SCR is the reaction between the NOx in exhaust gases and the reducing agent (NH_3 as ammonia water or urea solution) at the so-called deNOx catalyst at temperatures below 400 °C to produce N_2 and water vapor. The reaction formulas for NO and NOx are as follows [12,14]:

$$4NO + 4NH_3 + O_2 \rightarrow 4N_2 + 6H_2O$$

$$4NH_3 + 6NO \rightarrow 5N_2 + 6H_2O$$

$$2NO_2 + 4NH_3 + O_2 \rightarrow 3N_2 + 6H_2O$$

$$8NH_3 + 6NO_2 \rightarrow 7N_2 + 6H_2O$$

Accordingly, SCR needs twice as much NH_3 for NO than for NO_2 reduction.

The design of the supporting material of the SCR catalysts (TiO_2, ZrO_2, Al_2O_3, and ZnO) and the synthesis method of the system affect a final catalyst structure not only in the direct titania phase character, e.g., surface area and porosity structure, but also indirectly by deciding the structure of the W and V surface deposits, which form the acid catalytic centers controlling the mechanisms of the SCR reaction with NH_3 [14]. Specifically, NOx needs to be removed from the flue gases, particularly in electric power stations where generally TiO_2-SCR catalysts are used. Regenerating the spent TiO_2-SCR catalysts is a significant problem that needs further improvement despite many available options [15,16]. Currently, industrial TiO_2-based SCR systems suffer from surface deposits. Moreover, the exploitation deteriorates surface texture. As SCR catalysts are expensive, they need to be regenerated. The management of deactivated SCR catalysts should minimize the adverse environmental effects of these materials. On the other hand, it could also be a valuable resource of rare chemical elements such as vanadium and tungsten [17].

In particular, we focused this study on nitrogen dioxide (NO_2) mitigation. NO_2 is the most toxic NOx form in the atmosphere. Both indoor and outdoor NO_2 pollution exposure to humans was extensively studied [18]. At the same time, there are only several publications reporting the catalytic decomposition of NO_2, even though a complex NO/NO_2 SCR should support a standard NO SCR [19,20]. Recently, catalytic decomposition of NO_2 over a copper-decorated metal-organic framework by non-thermal plasma was studied [21].

The industrial SCR installations are TiO_2 layers with surface-engineered coatings by metal oxides. Zinc oxide is a low-cost TiO_2 potential alternative for catalytic SCR applications in environmental catalysis. However, the reported applications in this area, particularly NOx removal, are rare. The low ZnO stability is a reason why titania is much more popular. For example, the surface of polar ZnO nanoparticles undergoes significant change during storage at room temperature in the presence of moisture, oxygen, or light. During two-month storage, the specific surface area of the ZnO decreases from 115 to 35 m^2 g by room temperature sintering of polar ZnO nanosheets [22]. Recent efforts involve the potential enhancement of ZnO stability by structural modifications, such as doping [23] and core-shell nanoparticle formation [24], e.g., using hybrid ZnO-TiO$_2$ systems [25]. Surface deposits, e.g., Cu$_2$O/MoS$_2$/ZnO composites on Cu mesh, can upgrade the ZnO system, reducing N$_2$ to NH$_3$ (with water as the proton source) in the liquid membrane reactor under simulated visible light [26]. The room temperature sintering of polar ZnO nanosheets was prevented by forming the surface layer of silica (2 atom %) [27]. The engineering of ZnO-structured surfaces is an issue of general interest. In particular, nanostructures (nanowires, nanotubes, nanobelts, nanorings, flower-like morphologies, multipods, tetrapods, and sponge-like structures) were broadly investigated [28]. Chemical and Physical Vapor Deposition (CVD and PVD) sputtering, as well as evaporation approaches and epitaxial growth, are the methods for forming the continuous functional thin layers on ZnO. Wet chemical and template-assisted methods are alternatives [28].

For the first time, here, we investigated the influence of the ornamentation of Al$_2$O$_3$, SiC, and ZrO$_2$ ceramic foams by nanorod ZnO and TiO$_2$ coatings as W and V support on DeNOx catalysis, in particular on the NO$_2$ SCR process (gas flow rate of 3 dm^3/h and temperature of 400 °C at atmospheric pressure). In particular, we compared the broad library of the VOx or WOx supported on the SiC, Al$_2$O$_3$, ZnO, CeO$_2$, MgO, SiO$_2$, TiO$_2$, and ZrO$_2$ in the DeNOx within the temperature range of 200–400 °C. Physical and chemical analyses of the functionalized foam surface confronted with the catalyst performance indicate that these materials can be a new efficient SCR reaction platform.

2. Materials and Methods

2.1. Materials

Commercially available chemical reagents were used in the study: Tungsten powder <12 µm, 99.9% trace metal basis (Sigma-Aldrich, St. Louis, MO, USA); Vanadium powder 100 mesh, 99.9% trace metal basis (Sigma-Aldrich); and 30% hydrogen peroxide (Avantor Performance Materials., Gliwice, Poland). All chemicals were used without further purification.

2.2. Preparation of TiO$_2$ or ZnO Nanofilaments at Ceramic Foams

ZnO nanorods were prepared according to the following procedure. First, we used the ALD method to deposit the zinc oxide nanoseeds on Al$_2$O$_3$, SiC, and ZrO$_2$ substrates. We performed 12 ALD cycles using diethylzinc (DEZ; Sigma-Aldrich) as a zinc precursor and deionized water as an oxygen precursor at the temperature of 100 °C in the ALD Savannah 100 reactor from Cambridge NanoTech, Waltham, MA, USA (now called the Vecco Savanah® Series). The hydrothermal process of zinc oxide nanorods' growth consists of two steps, as described previously [29]. First, the reaction mixture with a Zn concentration of 1 mM was prepared. Zinc acetate dihydrate (Roth, 99% pure) was dissolved in 60 mL of deionized water. Afterward, the pH of the solution was adjusted to 7.5 using a 1 M aqueous solution of sodium hydroxide (Sigma-Aldrich). The prepared mixture was heated using an induction heater at 95 °C with nucleated substrates inside and kept at this temperature for approximately 2 min. Then, the samples were removed from the reactor, rinsed with isopropanol, and dried in air.

Part of the received substrates with ZnO nanorods were used as a scaffold for the growth of the nanostructured TiO$_2$ surface. These substrates covered by ZnO nanorods were placed in the ALD reactor at 100 °C. The 90 ALD cycles deposited approximately 50 nm thick layers of TiO$_2$. We used tetrakis (dimethylamino) titanium (IV; TDMAT) from STREM Chemicals as a titanium precursor and deionized water as the oxygen precursor. After TiO$_2$ deposition, nanorods were etched using 1% HCl acid for 1 min, rinsed in water, and dried.

2.3. General Preparation of W and V Nanoparticles at Ceramic Foams and Other Carriers

To a weight of tungsten and/or vanadium powders (Table S1), 2 mL of 30% hydrogen peroxide solution was added and mixed for 3 h until the metal wholly dissolved. Ceramic foams described in paragraph 2.2 were powdered in mortar and sieved, and 990 mg of the selected material (Al_2O_3, ZrO_2, or SiC carrier with TiO_2 or ZnO_2 nanofilaments) was suspended in a solution containing tungsten and/or vanadium. The mixture was stirred until the solvent evaporated, yielding a powder catalyst. The catalysts containing tungsten and vanadium oxides deposited on powders ZnO, CeO_2, MgO, SiO_2, or TiO_2 (Sigma-Aldrich) were prepared using the same method (Table S2).

2.4. Methods of Catalyst Characterization

XPS measurements were completed using the PHI 5700 photoelectron spectrometer (Physical Electronics Inc., Chanhassen, MN, USA). Photoelectrons were excited by a monochromatic X-ray beam (Al Kα of energy of 1486.6 eV) from the sample surface. The resulting photoelectron spectra obtained for each element constituting the sample were analyzed using PHI MultiPak (v.9.6.0.15, ULVAC-PHI, Chigasaki, Japan) software. The obtained high-resolution spectra were calibrated using the C1s peak (284.8 eV), the occurrence of which is related to the presence of adsorbed carbon on the sample surface. Analyzed core levels were fitted using a combination Gauss–Lorentz shape of the photoemission line and Shirley background.

We used an energy-dispersive X-ray fluorescence (EDXRF) spectrometer—Epsilon 3 (Panalytical, Almelo, The Netherlands) to perform chemical analysis. The spectrometer was equipped with a thermoelectrically cooled silicon drift detector (SDD) and Rh target X-ray tube. It was operated at a maximum voltage of 30 keV and maximum power of 9 W. We applied the Omnian software with the fundamental parameter method for quantitative analysis. Measurement conditions were as follows: counting time 5 kV, 300 s, and helium atmosphere for Si and Al determination; counting time 12 kV, 300 s, air atmosphere, and 50 µm Al primary beam filter for V; 20 kV, counting time 120 s, air atmosphere, and 200 µm Al primary beam filter for Fe; and 30 kV, counting time 120 s, air atmosphere, and 100 µm Ag primary beam filter for W, Zn, Zr, Y, and Hf. We fixed the current of the X-ray tube not to exceed the dead-time loss of ca. 50%.

We used Hitachi SU-70 equipment (15 kV of accelerating voltage using a secondary electron detector) for Scanning Electron Microscopy (SEM) and the transmission electron microscopy (TEM) was performed in the JEOL high-resolution (HRTEM) JEM 3010 microscope operating at a 300 kV accelerating voltage with a Gatan 2k × 2k OriusTM 833SC200D CCD camera and an EDS detector from IXRF Systems. The samples were suspended in isopropanol and deposited on a Cu grid with an amorphous carbon film standardized for TEM observations. Selected Area Electron Diffraction (SEAD) patterns were indexed using dedicated ElDyf software (Institute of Material Science, University of Silesia., Katowice, Poland).

The X-ray powder diffraction (XRD) measurements were carried out using a Malvern Panalytical Empyrean diffractometer. Cu anodes operated at a wavelength of 1.54056 Å, at an electric current of 30 mA and voltage of 40 kV, and equipped with a PIXcell3D solid-state hybrid pixel detector. The XRD was registered in the angular range of $2\theta = 15–145°$ with 0.02° steps. The phase analysis involved reference standards from the International Centre for Diffraction Data (ICDD) PDF-4 database. Rietveld refinement was performed using FullProf computer software (available at www.ill.eu/sites/fullprof/ (accessed on 10 February 2022)).

2.5. NOx Decomposition in a Flow Reactor

We used a fixed-bed quartz flow reactor using a 200 mg sample of the catalysts at 200–400 °C under atmospheric pressure to test the SCR catalysis performance. We crushed the catalyst to a fine powder before SCR tests. The feed gas was composed as follows I: 0.2% NO_2 + 5% O_2 + 94.8% He and inlet II: 0.2% NH_3 + 5% O_2 + 94.8% He (volume ratio 3:4). The tests were performed at the flow rate of 3 dm^3/h. We used the GC-FID method to monitor the gas composition. The NOx conversion amounted to:

$$NOx\ conversion = [(NOx\ inlet - NOx\ outlet)/NOx\ inlet] \times 100\%$$

3. Results and Discussion

3.1. The Catalysts Design, Preparation, and Structure

The exemplary EDXRF spectra of the selected catalysts collected using the Rh X-ray tube operating at 30 and 5 kV are given in Figure 1. Spectra of ZnO/Al$_2$O$_3$ show peaks at 6.40, 7.06, 8.64, 9.57, 15.77, 20.21, 22.72, 2.70, 1.49, and 1.74 keV, corresponding to Fe Kα, Fe Kβ, Zn Kα, Zn Kβ, Zr Kα, Rh Kα, Rh Kβ, Rh Lα (X-ray tube), Al Kα, and Si Kα, respectively. Spectra of ZnO/ZrO$_2$ show peaks at 2.04, 8.64, 9.57, 14.96, 15.77, and 17.67 keV, corresponding to Zr Kα, Zn Kα, Zn Kβ, Y Kα, Zr Kα, and Zr Kβ, respectively. The spectra also reveal the presence of Hf as common impurities of zirconium compounds (Lα, Lβ, and Lγ lines at 7.90, 9.02, and 10.52 keV, respectively). The EDXRF spectra of the 1.0% W,V/ZnO/Al$_2$O$_3$ and 1.0% W,V/ZnO/ZrO$_2$ systems reveal the presence of tungsten (Lα, Lβ, and Lγ lines at 8.40, 9.67, and 11.29 keV, respectively) and vanadium (Kα and Kβ lines at 4.95 and 5.43 keV, respectively). Table S3 presents the results of the quantitative EDXRF analysis of ZnO/Al$_2$O$_3$, 1.0% W/ZnO/Al$_2$O$_3$, 1.0% W,V/ZnO/Al$_2$O$_3$, ZnO/ZrO$_2$, 1.0% W/ZnO/ZrO$_2$, and 1.0% W,V/ZnO/ZrO$_2$.

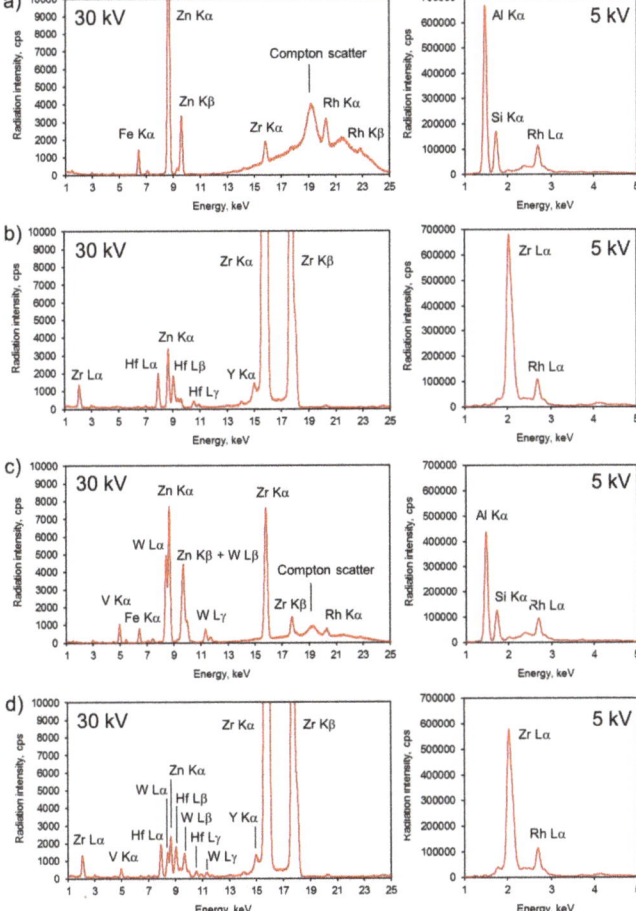

Figure 1. EDXRF spectrum of (**a**) reference sample—ZnO/Al$_2$O$_3$, (**b**) reference sample—ZnO/ZrO$_2$, (**c**) 1.0% W,V/ZnO/Al$_2$O$_3$ system, and (**d**) 1.0% W,V/ZnO/ZrO$_2$ system.

We performed X-ray powder diffraction measurements to examine the phase composition of the material samples. Compared with reference standards from the ICDD PDF4+ database of the Al_2O_3/ZnO sample, this analysis revealed four phases. The Al_2O_3 (PDF 04-006-9730) and Al_2SiO_5 (PDF 01-088-0892) phases were dominating, while we also observed the ZnO (PDF 01-078-4603) and SiO_2 (PDF 04-013-9484) phases. The additional nano-size SiO_2 phase (PDF 01-073-3436) was detected for the samples containing W or the combined W and V oxides. No additional phases nor impurities were observed. The performed full pattern Rietveld refinement allowed for the determination of the crystallographic parameters of the phases formed. Accordingly, the mean crystallite size based on the peak broadening was calculated. The obtained results of the Rietveld refinement are shown in Table S4. The X-ray powder diffraction measurements were also conducted for ZrO_2/ZnO samples. The phase analysis revealed the presence of two ZrO_2 phases. The dominant, monoclinic one (PDF 04-010-6452) was accompanied by the cubic phase (PDF 01-078-3193). Additionally, ZnO (PDF 01-078-4603) and Zn_2SiO_4 (PDF 01-076-8176) phases were also detected. Table S4 and Figure 2 show the results of the Rietveld refinement. We should remember that metallic oxides could not be observed with the X-ray powder diffraction method due to the X-ray diffraction detection limit. Accordingly, the XRF, XPS, and TEM techniques confirmed the presence of W and V oxides.

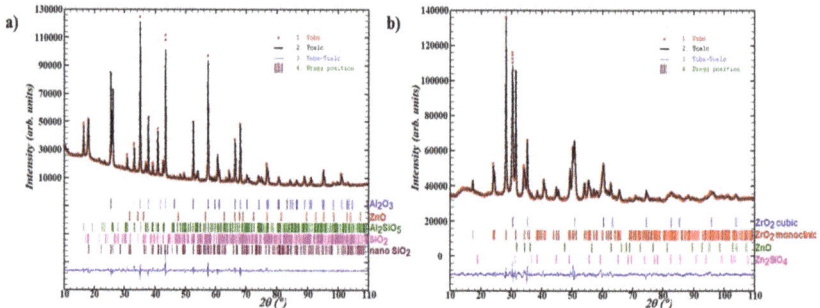

Figure 2. Rietveld refinement of the ZnO/Al_2O_3 (a) and ZnO/ZrO_2 (b) samples with identified phases. Reference dots indicate measurements points (Imes); the black solid curve calculated the pattern (Icalc); and the solid blue line indicates the (Imes-Icalc) difference, whereas vertical bars indicate Bragg position for the identified phases.

The TEM with an energy-dispersive X-ray detector confirmed the occurrence of metallic particles. The metallic nanoparticles occurred on the ceramic Al_2O_3 microparticles as indicated by the bright-field images (Figure 3). These structures were also proved with other spectroscopic measurements using XRF or XPS techniques that map the presence of metallic elements.

The XPS analysis determined the chemical states of the elements composing the samples. The main focus of the analysis was to determine the chemical states of vanadium and tungsten. We performed deconvolution of the V2p, W4f, and W4d photoemission lines. Additionally, lines O1s and C1s, and those associated with the Al_2O_3, ZrO_2, and ZnO matrix were analyzed.

The surface XPS spectra of the V2p core level, as shown in Figure 4a, indicate vanadium oxide in the studied samples. A slight shift in the position of the $V2p_{3/2}$ line was observed between examined systems; for the 1.0% W,V/ZnO/ZrO_2 system, the binding energy of the $V2p_{3/2}$ line is 516.96 eV, while for the 1.0% W,V/ZnO/Al_2O_3 system, it is 516.64 eV. The reference $V2p_{3/2}$ binding energies of the V_2O_5, as given in the NIST database [30], range from 516.6 eV to 517.7 eV. However, literature data specify that the 516.94 eV peak can also be assigned to vanadium 4+ [31]. The vanadium line is presented in Figure 4a together with the oxygen line. The location of the deconvoluted oxygen lines was assigned to the metal oxides or C=O bond, as also seen with the carbon C1s lines (not shown here).

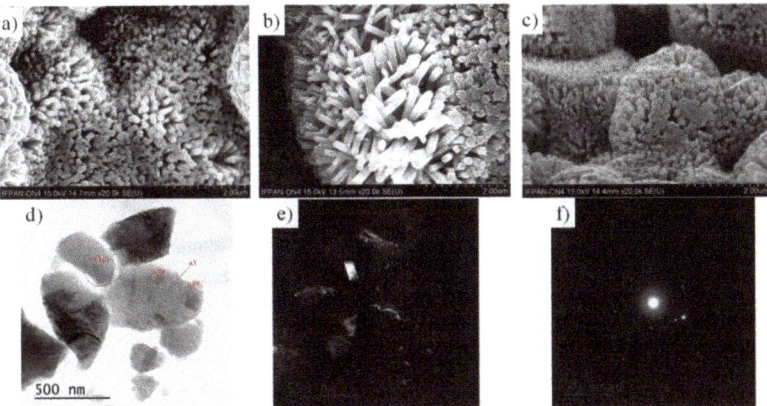

Figure 3. SEM micrographs of ZnO nanowires deposited on (**a**) SiC; (**b**) ZrO$_2$; and (**c**) Al$_2$O$_3$. TEM micrographs of V nanoparticles powdered 1.0% W,V/ZnO/Al$_2$O$_3$ catalyst (**d**,**e**) show recorded bright and dark field images, and (**f**) present recorded selected area electron diffraction patterns from regions shown in part (**d**).

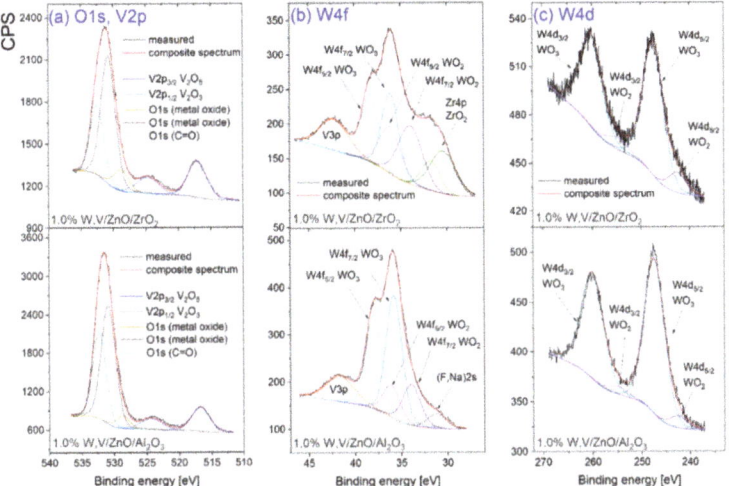

Figure 4. High-resolution XPS spectra of (**a**) O1s and V2p, (**b**) W4f, and (**c**) W4d. The top spectra represent 1.0% W,V/ZnO/ZrO$_2$ system and the spectra at the bottom represent the 1.0% W,V/ZnO/Al$_2$O$_3$ system.

The chemical state analysis of tungsten was based on two lines, namely W4f and W4d (see Figure 4b,c). For the W4f line, a superposition of the core levels originating from other elements detected on the sample surface was observed. Therefore, the tungsten chemical states, as identified through the W4f line analysis, were further examined, inspecting the shape of the W4d line.

The deconvoluted peaks of the W4f XPS spectra showed two oxides for both examined systems; the binding energy of the W4f$_{7/2}$ line at 33.79 eV can be assigned to WO$_2$ [32], whereas the peak at 35.76 eV can be assigned to the WO$_3$ [33]. Other peaks seen in the W4f line are associated with various elements detected on the sample's surface (e.g., Na and F were visible in the overview spectra, while Zr and V are components of the studied systems), as marked in Figure 4b. The presence of WO$_2$ and WO$_3$ oxides was confirmed by analysis of line W4d, as illustrated in Figure 4c. Analysis of the W4d line allows us to more accurately see the differences in the ratio of each oxide's contribution to a given system. XPS chemical analysis showed no significant differences in the spectra of the base compounds. The 1.0% W,V/ZnO/Al_2O_3 system and the corresponding reference sample showed the presence of the Al_2O_3 (Al2p line at 74.62 eV [34]). The binding energy of $Zn2p_{3/2}$ can be ascribed to the Zn^{2+} state [35] and ZnO oxide [36]; those chemical states were observed at 1020 eV and 1022.3 eV, respectively. Similarly, for the 1.0% W,V/ZnO/ZrO_2 system, the presence of zinc and zirconium oxides (ZnO assigned for $Zn2p_{3/2}$ at 1022.3 eV [36], and clusters of ZrO_2 for $Zr3d_{5/2}$ at 182.44 eV [37]) were detected. Additionally, a relatively small amount of zirconium oxide nanocrystallites for the $Zr3d_{5/3}$ at 181.2 eV [38]) was present on the sample surface. The positions of the photoemission lines were consistent with those observed for the reference sample. A detailed comparison of the XPS spectra for the 1.0% W,V/ZnO/ZrO_2 vs. 1.0% W,V/ZnO/Al_2O_3 vs. 1.0% W,V/ZnO/SiC systems is shown in Figure 4 and Figure S1 Supplementary Materials.

3.2. Catalyst Performance in SCR Reaction

We designed a broad library of tested catalyst systems by combining SiC, Al_2O_3, and ZrO_2 (foams) with ZnO, CeO_2, MgO, and SiO_2 or TiO_2 (coatings) with W and V oxides' loads. In these initial experiments, we decided to use the 1:1 W to V molar ratio due to the synergistic effect of W and V oxides in SCR catalysis [39,40]. A library of potential supports and coatings were pretested at five operating temperatures, assuming the maximum process temperature of 400 °C. Table S2 summarizes the results.

The NOx conversion was 22.5% vs. 19.7% vs. 18.1% at 250 °C for the most active systems Al_2O_3, MgO, and ZrO_2 (Table S2, entries 2, 5, and 8). The increase in temperature to 300 °C resulted in a slight increase in the conversion to 45.3% vs. 36.4% vs. 31.8%. However, it was not but at the temperature of 350 °C where a significant increase in the NOx conversion was observed, especially for ZrO_2 and Al_2O_3, where conversion takes a value of 85.2% and 80.6%, respectively (Table S2, entry 8 vs. 2). Increasing the temperature to 400 °C in the tested systems resulted in a slight decrease (ZnO, MgO, and ZrO_2) or increase (SiC, Al_2O_3, CeO_2, SiO_2, and TiO_2) of the NOx conversion rate (Table S2, entries 3, 5, and 8 vs. 1, 2, 4, 6, and 7). In turn, the W and V oxides supported by CeO_2, SiO_2, MgO, TiO_2, and SiC allowed for a relatively low conversion of NOx (60–79%) at 400 °C. Interestingly, the literature often describes these systems as an attractive alternative for commercial SCR catalysts [40–44]. In the context of the potential supporting foams (Al_2O_3, ZrO_2, and SiC), the Al_2O_3 and ZrO_2 outperform the SiC one. The catalytic performance of both Al_2O_3 and ZrO_2 is higher than 80% at 400 °C, which locates these supports just after the superior ZnO support (entries 2 vs. 8 vs. 3, Table S2).

As for Al_2O_3, considering the ZrO_2 foam supports appeared comparable in the catalytic performance tests (Table S2), we selected the Al_2O_3 support coated with nanorod ZnO for extensive and thorough testing. TiO_2 was selected as a comparison, providing illustrative insight into the current SCR systems. In particular, we tested the influence of the VOx to WOx ratio on the NOx conversion. This ratio was set at 1:0; 7:3; 1:1; 3:7; and 0:1, respectively. The nominal content of the supported metal oxides in all catalyst systems was 1%. We used the TiO_2 or ZnO nanorods at the Al_2O_3 carrier as the comparison standards (entries 1 and 7, Table 1).

For TiO_2-based systems with a higher VOx content vs. WO_3 (1:0 and 7:3, respectively), NOx conversion is slightly higher by 1.8–2.4% than the analogous ZnO-based systems (Table 1, entries 2 and 3 vs. 8 and 9). The opposite tendency was observed for the systems with VOx to WO_3 ratios of 1:1, 3:7, and 0:1 (Table 1, entries 4–6 vs. 10–12). The screen of the systems in Table 1 allowed us to optimize the molar ratio of the WOx to VOx. The best catalysts were selected based on the TON parameter. For ZnO/Al_2O_3 nanorods, the highest TON of 623 mmol/kg·h was observed for 1% V,W(3:7)/ZnO/Al_2O_3, while for TiO_2/Al_2O_3 nanorods, the highest TON of 623 mmol/kg·h was observed for the 1% W/TiO_2/Al_2O_3 (TON = 613 mmol/kg·h). Both for the TiO_2/Al_2O_3 nanorods and ZnO/Al_2O_3, we observed a systematic increase in TON when the VOx content decreased in favor of the WOx content (Table 1: 543 vs. 619 mmol/kg·h or 527 vs. 613 mmol/kg·h). Accordingly, in the next step, we prepared the ceramic foams with nanorod TiO_2 or ZnO coating with a VOx to WO_3 load at the optimal molar ratio of 3:7 (Table 1).

Table 1. The V and W load optimization on TiO_2 and ZnO nanorod-coated Al_2O_3 foam.

Entry	Catalyst	NOx Conversion [%] [a]	TON [$\frac{mmol}{kg \cdot h}$]
1	TiO_2/Al_2O_3	78.7	527
2	1% V/TiO_2/Al_2O_3	85.5	573
3	1% V,W(7:3)/TiO_2/Al_2O_3	89.4	599
4	1% V,W(1:1)/TiO_2/Al_2O_3	88.3	591
5	1% V,W(3:7)/TiO_2/Al_2O_3	90.2	604
6	1% W/TiO_2/Al_2O_3	91.5	613
7	ZnO/Al_2O_3	81.1	543
8	1% V/ZnO/Al_2O_3	82.9	555
9	1% V,W(7:3)/ZnO/Al_2O_3	87.6	587
10	1% V,W(1:1)/ZnO/Al_2O_3	90.8	608
11	1% V,W(3:7)/ZnO/Al_2O_3	93.0	623
12	1% W/ZnO/Al_2O_3	92.4	619

[a] Gas flow rate of 3 dm^3/h, temperature 400 °C, and atmospheric pressure.

Table 2 shows a detailed comparison of NOx conversions for ZnO nanofilaments with different surface coatings on ZrO_2, Al_2O_3, and SiC foam supports measured in SCR reaction. ZnO nanofilaments in the presence of VOx and WOx on the Al_2O_3 carrier turned out to be the most active (NOx conversion of 94.8%), slightly outperforming other carriers (ZrO_2 and SiC) with ZnO nanofilaments decorated with surface VOx and WOx (having a NOx conversion of 93.0% and 89.1%, respectively). Specifically, ZnO nanofilaments with VOx and/or WOx increase the deNOx activity of all crude foam supports. High activity of these systems may result from expanding the surface of the support by the nanorod coating.

Table 2. Catalytic performance of the Al_2O_3, SiC, and ZrO_2 foam coated with zinc oxide nanorods with surface loadings of V and W oxides.

Entry	Catalyst	Sample Code	NOx Conversion (%) [a]
1	ZnO/Al_2O_3	ZA	88.6
2	1% W/ZnO/Al_2O_3	W/ZA	90.3
3	1% V/ZnO/Al_2O_3	V/ZA	88.2
4	1% W,V/ZnO/Al_2O_3	W,V/ZA	94.8
5	ZnO/SiC	ZS	81.7
6	1% W/ZnO/SiC	W/ZS	89.9
7	1% V/ZnO/SiC	V/ZS	83.2
8	1% W,V/ZnO/SiC	W,V/ZS	89.1
9	ZnO/ZrO_2	ZZ	91.3
10	1% W/ZnO/ZrO_2	W/ZZ	92.4
11	1% V/ZnO/ZrO_2	V/ZZ	85.4
12	1% W,V/ZnO/ZrO_2	W,V/ZZ	93.0

[a] Gas flow rate of 3 dm^3/h, temperature 400 °C, and atmospheric pressure.

Figure 5 illustrates the catalytic performance of these systems in the temperature range of 200–400 °C. For the catalytic systems with TiO_2 nanorods, the highest degree of NOx conversion at 400 °C (88–96%) was noted when the carrier was ZrO_2 (Figure 5A), while the lowest was noted for SiC support (55–63%). ZnO nanorods deposited on Al_2O_3 made it possible to obtain the 90–95% degree of NOx conversion at 400 °C, while the worst carrier for ZnO nanorods was again SiC, with an 83–90% NOx conversion (Figure 5B). Generally, the comparison of TiO$_2$ and ZnO nanorods in Figure 5A,B reveals that the SCR conversions on ZnO appeared better than on the TiO$_2$ supports in the range of the tested temperatures.

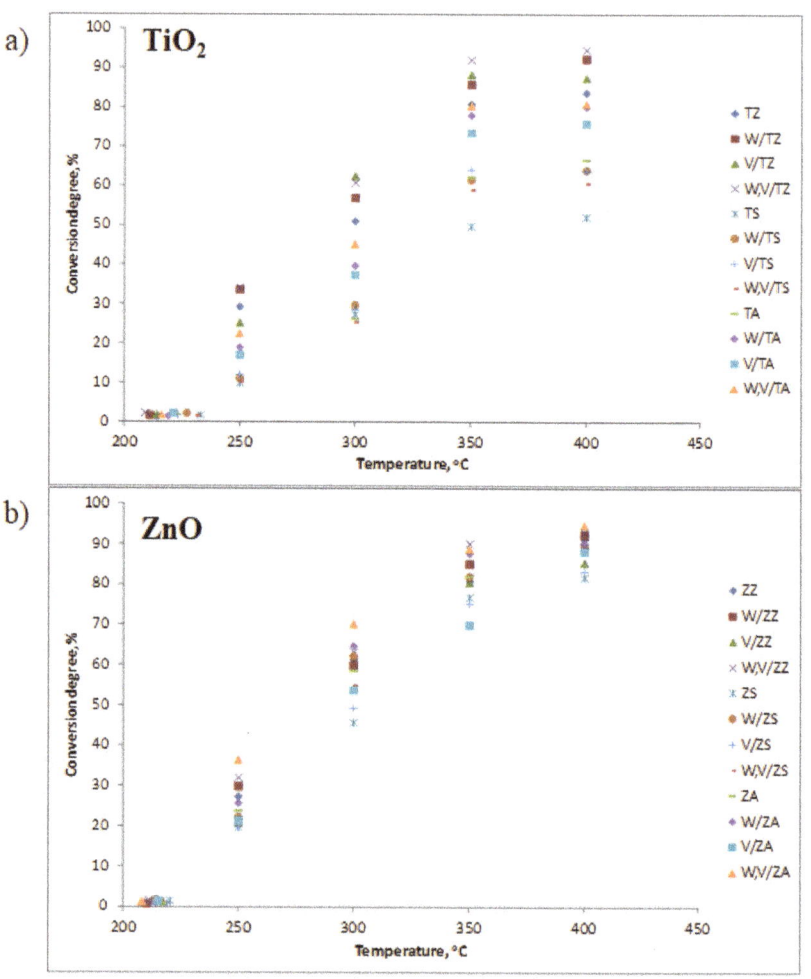

Figure 5. Catalytic activity of the powdered catalyst foam with (**a**) TiO$_2$ and (**b**) ZnO nanorods. Gas flow rate of 3 dm^3/h, temperature 400 °C, and atmospheric pressure. Acronyms: TZ—TiO$_2$/ZrO$_2$, W/TZ—1% W/TiO$_2$/ZrO$_2$, V/TZ—1% V/TiO$_2$/ZrO$_2$, W,V/TZ—1% W,V/TiO$_2$/ZrO$_2$, TS—TiO$_2$/SiC, W/TS—1% W/TiO$_2$/SiC, V/TS—1% V/TiO$_2$/SiC, W,V/TS—1% W,V/TiO$_2$/SiC, TA—TiO$_2$/Al$_2$O$_3$, W/TA—1% W/TiO$_2$/Al$_2$O$_3$, V/TA—1% V/TiO$_2$/Al$_2$O$_3$, W,V/TA—1% W,V/TiO$_2$/Al$_2$O$_3$, ZZ—ZnO/ZrO$_2$, W/ZZ—1% W/ZnO/ZrO$_2$, V/ZZ—1% V/ZnO/ZrO$_2$, W,V/ZZ—1% W,V/ZnO/ZrO$_2$, ZS—ZnO/SiC, W/ZS—1% W/ZnO/SiC, V/ZS—1% V/ZnO/SiC, W,V/ZS—1% W,V/ZnO/SiC, ZA—ZnO/Al$_2$O$_3$, W/ZA—1% W/ZnO/Al$_2$O$_3$, V/ZA—1% V/ZnO/Al$_2$O$_3$, and W,V/ZA—1% W,V/ZnO/Al$_2$O$_3$.

A comparison of the foam supports indicates that for the oxide-supported (Al$_2$O$_3$ and ZrO$_2$) catalysts, SCR reactivity slightly outperforms the SiC supported systems. Formally, all supports are semiconductors of the wide gap of ca. 3.2 ÷ 3.37 eV, but in nano ZnO or TiO$_2$, the gap energy can be modulated by nanostructure organization and interactions with supported metals (W or V) present at the surface in the form of metal oxides. Optical spectra of the ZnO and TiO$_2$ supported on Al$_2$O$_3$ and SiC (Figure S2, Supplementary Materials) indicate a correlation between deNOx behavior and the energy band gap. In particular, unlike the Al$_2$O$_3$, the SiC-supported ZnO and TiO$_2$ systems indicate an additional low systems energy absorption band. It is, however, not clear if this can affect the thermal (dark) SCR reaction.

Interestingly, material structure and coatings can influence the band gap value. For example, the tungsten oxides' band gap can be reduced to 2.47 through the interactions with other semiconductors, e.g., CdTe [45]. Similarly, the contacts of Ti and V$_2$O$_5$, even as large crystallites, modified the optical band gap (1.96 eV vs. 2.2 eV for undoped V$_2$O$_5$). This effect is attributed to lattice expansion by the Ti ion and oxygen vacancies formation [46]. The specific interactions of the individual nanorod coating and synergistic interaction of the V and W oxide coatings can also play a role. For example, the doping of the W oxide with V can result in synergic interactions downshifting the material band gap [47]. In turn, the band-gap energies of the SiC nanowires are higher than the corresponding SiC bulk values [48].

4. Conclusions

Energy production is associated with many environmental risks; e.g., generating greenhouse gas emissions and nitrogen oxides (NOx) are among the primary gas pollution concerns. Environmental catalysis plays a pivotal role in NOx mitigation (DeNOx). This study investigated a collection of ceramic foams as a potential catalyst support for selective catalytic NOx reduction (SCR). To be an attractive support, we should functionalize the surface of raw foams. A library of ceramic SiC, Al$_2$O$_3$, and ZrO$_2$ foams ornamented with nanorod ZnO and TiO$_2$ as W and V oxide coatings was obtained for the first time. We characterized the surface layer structure using XPS, XRF, and SEM and TEM microscopy to optimize the W to V molar ratio.

NOx generation, pollution, and reduction are complex problems. First, combustion in air yields two forms of NOx, namely NO and NO$_2$, in the ratio NO/NOx of 0.90 to 0.95. However, the dominating NOx form in the atmosphere is NO$_2$ resulting from NO oxidation. Accordingly, the main issue in the environmental catalysis of NOx refers to NO, while the main topics of NOx ecotoxicology refer to NO$_2$. Low NO$_2$ content in flue gases also decides NO$_2$. SCR is tested only very rarely, even though a complex NO/NO$_2$ SCR should support a standard NO SCR; therefore, here, we used the NO$_2$ SCR model.

Comparing TiO$_2$ and ZnO systems reveals that the SCR conversion on ZnO appeared superior vs. conversion on TiO$_2$, while the SiC-supported catalysts were less efficient than Al$_2$O$_3$ and ZrO$_2$-supported catalysts. The energy bands in optical spectra correlate with the observed activity rank. However, a more detailed study is needed to assess whether this effect is coincidental only.

Supplementary Materials: The following are available online at https://www.mdpi.com/article/10.3390/en15051798/s1, Table S1: The amount of metal needed to obtain vanadium, tungsten, or mixed catalysts calculated at 990 mg of the carrier; Table S2: NOx decomposition data for V and W oxides deposited on different carriers. Gas flow rate of 3 [dm^3/h]; Table S3: EDXRF analyses of Ti, W, and V nanoparticles deposited on powder or foam Al$_2$O$_3$, SiC, and ZrO$_2$; Table S4: The average crystallite size and lattice parameters of investigated materials determined by XRD method; Figure S1: High-resolution XPS spectra of (a) O1s and V2p, (b) C1s, (c) W4f, and (d) W4d. The red spectra represent 1.0% W,V/ZnO/ZrO$_2$ system, the blue—the 1.0% W,V/ZnO/Al$_2$O$_3$ system, and the green—1.0% W,V/ZnO/SiC system; and Figure S2: UV-Vis spectra of TiO2/SiC or TiO$_2$/Al$_2$O$_3$ and ZnO/SiC or ZnO/Al$_2$O$_3$ in powders.

Author Contributions: Conceptualization, M.K. and J.P.; formal analysis, J.P., M.K. and P.B.; writing—original draft preparation, M.K., J.P., T.S., P.B., R.S., M.Z., J.S., K.B., B.S.W., M.O., R.P. and M.G.; writing—review and editing, P.B. and M.K.; visualization, M.K.; supervision, J.P. and T.S.; project administration, J.P. and P.B.; funding acquisition, J.P. All authors have read and agreed to the published version of the manuscript.

Funding: This research study was funded by the National Centre for Research and Development in Poland, grant no. TANGO1/266384/NCBR/2015, and National Science Center, grant no. OPUS 2018/29/B/ST8/02303.

Institutional Review Board Statement: Not applicable.

Informed Consent Statement: Not applicable.

Data Availability Statement: The datasets generated during the current study are available from the corresponding authors upon reasonable request.

Acknowledgments: For the support, Jaroslaw Polanski would like to acknowledge Zielony Horyzont: New Energy project ZFIN 40001022.

Conflicts of Interest: The authors declare no conflict of interest.

References

1. Polanski, J.; Lach, D.; Kapkowski, M.; Bartczak, P.; Siudyga, T.; Smolinski, A. Ru and Ni—Privileged Metal Combination for Environmental Nanocatalysis. *Catalysts* **2020**, *10*, 992. [CrossRef]
2. Van Noyen, J.; Mullens, S.; Snijkers, F.; Luyten, J. Catalyst design with porous functional structures. *Sustain. Chem. WIT Trans. Ecol. Environ.* **2011**, *154*, 93–102. [CrossRef]
3. Carty, W.M.; Lednor, P.W. Monolithic ceramics and heterogeneous catalysts: Honeycombs and foams. *Curr. Opin. Solid State Mater. Sci.* **1996**, *1*, 88–95. [CrossRef]
4. Twigg, M.V.; Richardson, J.T. Theory and Applications of Ceramic Foam Catalysts. *Chem. Eng. Res. Des.* **2002**, *90*, 183–189. [CrossRef]
5. Wang, Y.; Yang, Q.; Ke, L.; Peng, Y.; Liu, Y.; Wu, Q.; Tian, X.; Dai, L.; Ruan, R.; Jiang, L. Review on the catalytic pyrolysis of waste oil for the production of renewable hydrocarbon fuels. *Fuel* **2021**, *283*, 119170. [CrossRef]
6. Palma, V.; Ruocco, C.; Cortese, M.; Martino, M. Recent Advances in Structured Catalysts Preparation and Use in Water-Gas Shift Reaction. *Catalysts* **2019**, *9*, 991. [CrossRef]
7. Rodríguez, E.M.; Rey, A.; Mena, E.; Beltrán, F.J. Application of solar photocatalytic ozonation in water treatment using supported TiO_2. *Appl. Catal. B Environ.* **2019**, *254*, 237–245. [CrossRef]
8. Leonzio, G. ANOVA analysis of an integrated membrane reactor for hydrogen production by methane steam reforming. *Int. J. Hydrog. Energy* **2019**, *44*, 11535–11545. [CrossRef]
9. Zuercher, S.; Pabst, K.; Schaub, G. Ceramic foams as structured catalyst inserts in gas–particle filters for gas reactions—Effect of backmixing. *Appl. Catal. A Gen.* **2009**, *357*, 85–92. [CrossRef]
10. Huangfu, L.; Abubakar, A.; Li, C.; Li, Y.; Wang, C.; Gao, S.; Liu, Z.; Yu, J. Development of Red Mud Coated Catalytic Filter for NOx Removal in the High Temperature Range of 300–450 °C. *Catal. Lett.* **2020**, *150*, 702–712. [CrossRef]
11. Seijger, G.B.F.; Oudshoorn, O.L.; Boekhorst, A.; van Bekkum, H.; van den Bleek, C.M.; Calis, H.P.A. Selective catalytic reduction of NO over zeolite-coated structured catalyst packings. *Chem. Eng. Sci.* **2001**, *56*, 849–857. [CrossRef]
12. Seo, H.J.; Jeong, R.H.; Boo, J.-H.; Song, J.; Boo, J.-H. Study on Chemical Removal of Nitric Oxide (NO) as a Main Cause of Fine Dust (Air Pollution) and Acid Rain. *Appl. Sci. Converg. Technol.* **2017**, *26*, 218–222. [CrossRef]
13. Irfan, M.F.; Goo, J.H.; Kim, S.D. Co_3O_4 based catalysts for NO oxidation and NOx reduction in fast SCR process. *Appl. Catal. B Environ.* **2008**, *78*, 267–274. [CrossRef]
14. Lai, J.K.; Wachs, I.E. A Perspective on the Selective Catalytic Reduction (SCR) of NO with NH_3 by Supported V_2O_5–WO_3/TiO_2 Catalysts. *ACS Catal.* **2018**, *8*, 6537–6551. [CrossRef]
15. Kapkowski, M.; Siudyga, T.; Sitko, R.; Niemczyk-Wojdyla, A.; Zelenka, T.; Zelenkova, G.; Golba, S.; Smolinski, A.; Polanski, J. Toward a viable ecological method for regenerating a commercial SCR catalyst–Selectively leaching surface deposits and reconstructing a pore landscape. *J. Clean. Prod.* **2021**, *316*, 128291. [CrossRef]
16. Zhang, Q.; Wu, Y.; Yuan, H. Recycling strategies of spent V_2O_5-WO_3/TiO_2 catalyst: A review. *Resour. Conserv. Recycl.* **2020**, *161*, 104983. [CrossRef]
17. Huo, Y.; Chang, Z.; Li, W.; Liu, S.; Dong, B. Reuse and valorization of vanadium and tungsten from waste V_2O_5–WO_3/TiO_2 SCR catalyst. *Waste Biomass Valorization* **2015**, *6*, 159–165. [CrossRef]
18. Salonen, H.; Salthammer, T.; Morawska, L. Human exposure to NO_2 in school and office indoor environments. *Environ. Int.* **2019**, *130*, 104887. [CrossRef]
19. Koebel, M.; Madia, G.; Raimondi, F.; Wokaun, A. How NO_2 Affects the Reaction Mechanism of the SCR Reaction. Available online: https://www.osti.gov/etdeweb/servlets/purl/20398744 (accessed on 10 February 2022).
20. Zhu, X.; Zhang, L.; Dong, Y.; Ma, C. NO_2–NH_3 SCR over Activated Carbon: A Combination of NH_4NO_3 Formation and Consumption. *Energy Fuels* **2021**, *35*, 6167–6178. [CrossRef]
21. Xu, S.; Han, X.; Ma, Y.; Duong, T.D.; Lin, L.; Gibson, E.K.; Sheveleva, A.; Chansai, S.; Walton, A.; Ngo, D.-T.; et al. Catalytic decomposition of NO_2 over a copper-decorated metal–organic framework by non-thermal plasma. *Cell Rep. Phys. Sci.* **2021**, *2*, 100349. [CrossRef]

22. Fernández-Pérez, A.; Rodríguez-Casado, V.; Valdés-Solís, T.; Marbán, G. Room temperature sintering of polar ZnO nanosheets: I-evidence. *Phys. Chem. Chem. Phys.* **2017**, *19*, 16406–16412. [CrossRef]
23. Colón, G.; Hidalgo, M.C.; Navío, J.A.; Melián, E.P.; Díaz, O.G.; Rodríguez, J.M.D. Highly photoactive ZnO by amine capping-assisted hydrothermal treatment. *Appl. Catal. B Environ.* **2008**, *83*, 30–38. [CrossRef]
24. Yermolayeva, Y.V.; Savin, Y.N.; Tolmachev, A.V. Controlled Growth of ZnO Nanocrystals on the Surface of SiO$_2$ Spheres. *Solid State Phenom.* **2009**, *151*, 264–268. [CrossRef]
25. Siwinska-Stefanska, K.; Kubiak, A.; Piasecki, A.; Goscianska, J.; Nowaczyk, G.; Jurga, S.; Jesionowski, T. TiO$_2$-ZnO Binary Oxide Systems: Comprehensive Characterization and Tests of Photocatalytic Activity. *Materials* **2018**, *11*, 841. [CrossRef] [PubMed]
26. Su, Q.; Wang, W.; Zhang, Z.; Duan, J. Enhanced photocatalytic performance of Cu$_2$O/MoS$_2$/ZnO composites on Cu mesh substrate for nitrogen reduction. *Nanotechnology* **2021**, *32*, 285706. [CrossRef]
27. Fernández-Pérez, A.; Marbán, G. Room temperature sintering of polar ZnO nanosheets: III-Prevention. *Microporous Mesoporous Mater.* **2020**, *294*, 109836. [CrossRef]
28. Laurenti, M.; Stassi, S.; Canavese, G.; Cauda, V. Surface Engineering of Nanostructured ZnO Surfaces. *Adv. Mater. Interfaces* **2017**, *4*, 1600758. [CrossRef]
29. Witkowski, B.S.; Wachnicki, L.; Gieraltowska, S.; Dluzewski, P.; Szczepanska, A.; Kaszewski, J.; Godlewski, M. Ultra-fast growth of the monocrystalline zinc oxide nanorods from the aqueous solution. *Int. J. Nanotechnol.* **2014**, *11*, 758–772. [CrossRef]
30. NIST X-ray Photoelectron Spectroscopy Database, Version 4.1 (National Institute of Standards and Technology, Gaithersburg, 2012). Available online: http://srdata.nist.gov/xps/ (accessed on 10 February 2022).
31. Kim, D.; Kim, M.; Yi, J.; Nam, S.H.; Boo, J.H.; Park, Y.S.; Lee, J. Growth and characterization of VO$_2$ thin film by pulsed DC sputtering of optical switching applications. *Sci. Adv. Mater.* **2017**, *9*, 1415–1419. [CrossRef]
32. Cheng, I.C.; Garcia-Sanchez, E.; Hodge, A.M. Note: A method for minimizing oxide formation during elevated temperature nanoindentation. *Rev. Sci. Instrum.* **2014**, *85*, 096106. [CrossRef]
33. Korduban, A.M.; Shpak, A.P.; Medvedskij, M.M. Electronic structure exploration of active element surface for hydrogen sensor based on WO$_{3-x}$ nanoparticles. In *Hydrogen Materials Science and Chemistry of Carbon Nanomaterials*; Springer: Dordrecht, The Netherlands, 2007; pp. 59–64. [CrossRef]
34. Mendialdua, J.; Casanova, R.; Rueda, F.; Rodríguez, A.; Quiñones, J.; Alarcón, L.; Escalante, E.; Hoffmann, P.; Taebi, I.; Jalowiecki, L. X-ray photoelectron spectroscopy studies of laterite standard reference material. *J. Mol. Catal. A Chem.* **2005**, *228*, 151–162. [CrossRef]
35. Zhang, M.; Qin, J.; Yu, P.; Zhang, B.; Ma, M.; Zhang, X.; Liu, R. Facile synthesis of a ZnO–BiOI p–n nano-heterojunction with excellent visible-light photocatalytic activity. *Beilstein J. Nanotechnol.* **2018**, *9*, 789–800. [CrossRef] [PubMed]
36. Archana, J.; Navaneethan, M.; Hayakawa, Y. Morphological transformation of ZnO nanoparticle to nanorods via solid-solid interaction at high temperature annealing and functional properties. *Scr. Mater.* **2016**, *113*, 163–166. [CrossRef]
37. Li, H.; Rameshan, C.; Bukhtiyarov, A.V.; Prosvirin, I.P.; Bukhtiyarov, V.I.; Rupprechter, G. CO$_2$ activation on ultrathin ZrO$_2$ film by H$_2$O co-adsorption: In situ NAP-XPS and IRAS studies. *Surf. Sci.* **2019**, *679*, 139–146. [CrossRef]
38. Tsunekawa, S.; Asami, K.; Ito, S.; Yashima, M.; Sugimoto, T. XPS study of the phase transition in pure zirconium oxide nanocrystallites. *Appl. Surf. Sci.* **2005**, *252*, 1651–1656. [CrossRef]
39. Burdeinaya, T.N.; Matyshak, V.A.; Tret'yakov, V.F.; Glebov, L.S.; Zakirova, A.G.; Garcia, M.C.; Villanueva, M.A. Design of catalysts for deNOx process using synergistic phenomenon. *Appl. Catal. B Environ.* **2007**, *70*, 128–137. [CrossRef]
40. Misono, M. Chemistry and Catalysis of Mixed Oxides. In *Studies in Surface Science and Catalysis*; Elsevier: Amsterdam, The Netherlands, 2013; Volume 176, pp. 25–65. [CrossRef]
41. Zhao, X.; Mao, L.; Dong, G. Mn-Ce-V-WOx/TiO$_2$ SCR Catalysts: Catalytic Activity, Stability and Interaction among Catalytic Oxides. *Catalysts* **2018**, *8*, 76. [CrossRef]
42. Chen, L.; Weng, D.; Si, Z.; Wu, X. Synergistic effect between ceria and tungsten oxide on WO$_3$–CeO$_2$–TiO$_2$ catalysts for NH$_3$-SCR reaction. *Prog. Nat. Sci.* **2012**, *22*, 265–272. [CrossRef]
43. Selleri, T.; Gramigni, F.; Nova, I.; Tronconi, E. NO oxidation on Fe- and Cu-zeolites mixed with BaO/Al$_2$O$_3$: Free oxidation regime and relevance for the NH$_3$-SCR chemistry at low temperature. *Appl. Catal. B Environ.* **2018**, *225*, 324–331. [CrossRef]
44. Ganjkhanlou, Y.; Janssens, T.V.W.; Vennestrøm, P.N.R.; Mino, L.; Paganini, M.C.; Signorile, M.; Bordiga, S.; Berlier, G. Location and activity of VOx species on TiO$_2$ particles for NH$_3$-SCR catalysis. *Appl. Catal. B Environ.* **2020**, *278*, 119337. [CrossRef]
45. Hendi, A.H.Y.; Al-Kuhaili, M.F.; Durrani, S.M.A.; Faiza, M.M.; Ul-Hamid, A.; Qurashi, A.; Khan, I. Modulation of the band gap of tungsten oxide thin films through mixing with cadmium telluride towards photovoltaic applications. *Mater. Res. Bull.* **2017**, *87*, 148–154. [CrossRef]
46. Srilakshmi, P.; Maheswari, A.U.; Sajeev, V.; Sivakumar, M. Tuning the optical bandgap of V$_2$O$_5$ nanoparticles by doping transition metal ions. *Mater. Today* **2019**, *18*, 1375–1379. [CrossRef]
47. Karuppasamy, K.M.; Subrahmanyam, A. Results on the electrochromic and photocatalytic properties of vanadium doped tungsten oxide thin films prepared by reactive dc magnetron sputtering technique. *J. Phys. D Appl. Phys.* **2008**, *41*, 035302. [CrossRef]
48. Oliveira, J.B.; Morbec, J.M.; Miwa, R.H. Mechanical and electronic properties of SiC nanowires: An ab initio study. *J. Appl. Phys.* **2017**, *121*, 104302. [CrossRef]

MDPI
St. Alban-Anlage 66
4052 Basel
Switzerland
Tel. +41 61 683 77 34
Fax +41 61 302 89 18
www.mdpi.com

Energies Editorial Office
E-mail: energies@mdpi.com
www.mdpi.com/journal/energies

www.ingramcontent.com/pod-product-compliance
Lightning Source LLC
LaVergne TN
LVHW070045120526
838202LV00101B/433